ADVENTURES IN
FLIGHT SIMULATOR

THE ULTIMATE DESKTOP PILOT'S GUIDE

PUBLISHED BY
Microsoft Press
A Division of Microsoft Corporation
One Microsoft Way
Redmond, Washington 98052-6399

Copyright © 1994 by Timothy Trimble

All rights reserved. No part of the contents of this book may be reproduced or transmitted in any form or by any means without the written permission of the publisher.

Library of Congress Cataloging-in-Publication Data

Trimble, Timothy, 1957–
   Adventures in Flight simulator / Timothy Trimble.
      p.   cm.
   Includes index.
   ISBN 1-55615-582-4
    1. Microsoft Flight simulator (Computer file)   I. Title.
TL712.5.T75   1994
794.8'753--dc20   93-46243
                                                          CIP

Printed and bound in the United States of America.

1 2 3 4 5 6 7 8 9   QEQE 9 8 7 6 5 4

Distributed to the book trade in Canada by Macmillan of Canada, a division of Canada Publishing Corporation.

A CIP catalogue record for this book is available from the British Library.

Microsoft Press books are available through booksellers and distributors worldwide. For further information about international editions, contact your local Microsoft Corporation office. Or contact Microsoft Press International directly at fax (206) 936-7329.

Flight Simulator is a trademark of Bruce A. Artwick. Microsoft, Microsoft Press, MS, and MS-DOS are registered trademarks of Microsoft Corporation.

Acquisitions Editors: Marjorie Schlaikjer, Lucinda Rowley
Project Editors: Mary Ann Jones, Nancy Siadek, Kathy Krause
Manuscript Editors: Mary Ann Jones, Alice Copp Smith
Technical Reviewer: Lynn Guthrie
Technical Editor: Bill Teel

*To my beautiful wife, Denise, and our children, who tolerated
my hours and my dedication to this book and to the Flight Simulator
product. They continue to love me, and I love them.*

*And to my flight instructor, Bill Clark,
whose laid-back style and professionalism have helped me discover
the joy of flying in the real world.*

Contents

Preface — ix
Acknowledgments — xiii
Introduction — xv

Chapter 1
Introductory Flight — 1

   The Preflight — 2
   Taxi Time — 10
   The Runup — 12
   Departure — 16
   We Have Liftoff! — 18
   Over Chicago — 24

Chapter 2
Pilot Training and Private Lessons — 27

   Want To Be a Pilot? — 28
   Springtime in Paris — 31
   Preparing for Landings — 46
   Land Ho! Landing the Aircraft — 58
   Pattern Flying — 64

Chapter 3
Weather — 73

   The Importance of Weather — 74
   Clouds — 75
   Head in the Clouds — 79

Everyone Knows It's Windy	84
Too Hot to Handle	88
I Can't Stand the Pressure!	89
Weather Areas	90

Chapter 4
Instrument Flight   99

Instrument School	100
Situational Awareness	117
Reno to Oakland	119

Chapter 5
Graphics, Scenery, Pictures, and Video   137

The Art of Using Art	138
Graphics	141
Scenery	150
Follow the Leader	157
Pictures	159
Video	160

Adventure 1
Radio-Controlled Flight — 164

Adventure 2
Corporate Pilot — 172

Adventure 3
Engine Out Over Innsbruck — 185

Adventure 4
The Pilot Who Terrorized Paris — 192

Adventure 5
Lost and Found — 199

Adventure 6
Leisure Flight — 203

Appendix
Recommended System Requirements — 215

Index — 217

Preface

I was a teenager the first time I flew an airplane, which was also the first time I was completely and thoroughly scared. Several of my friends were taking flying lessons and had been filling my head with the excitement, the thrills, and the strange feeling of peace that comes from soaring in an airplane. By the time I finally decided to give it a try, I had become completely familiar with the terminology of flying. As a result, my first instructor assumed that I had been flying before, and I, of course, had assumed that I could handle whatever the instructor wanted me to do. So why should I expose myself as a first-timer?

The first surprise came when the instructor decided to let *me* take off. My heart jumped into my throat and kept me from declining the offer.

"Just smoothly push the throttle all the way in, and use the rudder pedals to keep it centered." With these words, the instructor turned and calmly looked out his window, pretending not to watch my efforts.

My palms were sweaty and my pulse was racing as the Cessna 150 bounded into the air. When we reached training altitude, I was instructed on the method of scanning the horizon for the proper attitude of the aircraft. My awe soon replaced fear as I admired the beauty of the earth below and the small, puffy clouds just overhead. After learning straight and level flight and then turns and banks, my awe turned to pride as I realized that *I* was in control of the aircraft.

The instructor then interrupted my reverie with a few simple words. "All right, good work. Now let's try some stalls."

Stalls? Did he say *"stalls"*? My mind turned instantly to mush as I tried to grasp the concept of what a stall was. All of the sudden, my firm grasp of the terminology and dominance over the skill of flying turned into the stark fear of a teenage boy waiting for his life to end.

As far as my humble, earthbound knowledge of flying could tell me, a stall was something that happened when the plane would no longer be a flying object, but rather, a falling object. In a sense, I was correct. But my limited flying knowledge also told me that falling objects don't stop falling until they hit the ground.

Our first stall consisted of dropping the power and holding the nose of the plane up until it felt like someone kicked the chair out from under us. The screaming stall buzzer turned my stomach to jelly, and then I held on for dear life as the nose dropped toward the ground. I held the yoke and the throttle as the instructor coached me through applying full throttle again and then pulling the nose back up to straight and level flight.

What? You mean we don't have to die now? It was all I could do to keep my composure as we went through the "power-off" stall several more times. When it was time to fly back to the airport, I no longer admired the view but concentrated instead on how to breathe normally. The instructor had me follow his movements on the controls as we descended toward the runway, a descent that ended in a smooth and graceful landing "on the numbers."

I taxied back to the ramp under the direction of the instructor while silently thanking God that I was back on the ground. After we pulled into the parking spot, the instructor took me step-by-step through the power-off checklist and then opened his door. He mentioned that I did an excellent job and that he would get me a log book back at the line office. I opened my door, intending to follow him to the line office until I realized that my legs would not support my body weight. Remaining in my seat, I yelled back that I would be there in a few minutes. Yes! I hate to admit it, but I actually got weak in the knees from my first flying lesson. I managed to accumulate another 11 hours in the air before I decided that the expense was more than I could afford at the time.

Several years later, I had the good fortune to be bitten by the computer bug. So badly was I affected that I packed up my family and minimal belongings and headed off to California, the computer mecca at that time, and managed to land a programming position with one of the major motion picture studios. It was then that I was first exposed to Flight Simulator for the IBM PC. I was instantly enthralled at being able to "fly" again using a microcomputer. Oblivious to the laughter of coworkers, who would watch me shoot lunchtime ILS approaches while contorting my body and chair into the appropriate landing attitude, I continued to learn and fly with the help of this marvelous product.

Despite hearing "it's just a game" and "you can't learn to fly with that" from the skeptics who did not share my enthusiasm, I decided to push the capabilities of this product even further. After several years of working for the studios, I quit and went to work for a major software development company that was conveniently located close to a municipal airport. I wasn't particularly interested in flying the "real" thing, but I needed resources for my Flight Simulator package. After obtaining the local sectional maps and approach charts, I decided to push the newest version of Flight Simulator (FS 3 at the time) closer to being a real flight simulator. Because I was familiar with the Los Angeles basin area and had spent a lot of time watching the planes take off and land at Van Nuys airport (VNY), I decided to set up the ultimate situation. Using Flight Simulator, a Los Angeles sectional map, and an FAA-approved approach chart for Van Nuys, I set out to create the worst "white-knuckle" flight I could imagine. Using Flight Simulator 3, I created a low overcast ceiling of 1500 feet that topped out at 15,000 feet, thunderstorms, and 10-knot winds out of the west. The flight would be at night. If I could actually pull off this flight, my skills at navigation and IFR flight would be complete! Taking off from LAX (Los Angeles International), I climbed to 5500 feet and flew to the Fillmore VOR, and then flew away from the VOR along the 53° radial. By adhering to the navigational settings on the VNY approach chart, I intercepted the glide slope exactly where the chart said it would be. Adjusting for the wind and the proper descent rate was a bit tricky and gave me a good case of sweaty palms, but I managed to stay "on the needle" through the approach. Then came the greatest thrill of the whole flight—breaking through the clouds at 1500 feet and seeing the approach lights of the runway right in front of me. From that point on, I knew that Flight Simulator was a lot more than "just a toy."

I was convinced that a lot of real pilots were missing out on this economical way to keep their navigational skills sharp. With this thought in mind, I wrote my first Flight Simulator article, gearing it toward real pilots. The article was accepted by a popular flying magazine and was the stepping-stone for a regular column about flight simulation products in a computer gaming magazine.

My aim with this book is not only to entertain and inform but also to demonstrate that Flight Simulator 5 shrinks the gap between real flight and Flight Simulator. I can't help but think back to my weak knees after my first airplane flight. If I had been able to spend time with Flight Simulator before taking flying lessons, I would have been a lot more knowledgeable and proficient, and I would have had the confidence gained by understanding the characteristics of real flight.

I am currently putting into practice what I preach by finally completing my private pilot training with the assistance of Flight Simulator. The FAA won't let me count my Flight Simulator time as flight hours, but I am thoroughly convinced that when I find myself in a confusing navigational situation in the future, the skills that I have acquired with Flight Simulator will help me resolve my situation more efficiently than if I had never used this great product.

I hope that my enthusiasm is conveyed in this book and that you, as a real pilot or as an armchair pilot, gain the same satisfaction, enjoyment, and education from using this product that I have.

Acknowledgments

My sincere thanks go to the following individuals and companies for their help and support:

To Marjorie Schlaikjer, at Microsoft Press, for putting the book in motion and helping a nervous writer become comfortable with the concept of filling vast amounts of time by putting words to paper.

To Mary Ann Jones, Kathy Krause, Lucinda Rowley, and Nancy Siadek, at Microsoft Press, for adding grace and proper punctuation to these pages.

To Jon Solon, Stuart Burden, and Harry Emil, for their time and patience in answering my technical questions.

A special thanks to Bill Teel, at Microsoft Press, for his incredible attention to technical details and his enthusiasm. (Bill, you kept it fun!)

To my friends in Bethesda, for allowing me to spend my free hours away from home on their computers. (The next capo is on me!)

To Vista Flying Services, in Carlsbad, California, for seeing me through my first solo and for taking care of my favorite bird—63287.

To CH Products and Cathy Panos, for helping me to turn my computer into a lean, mean, flying machine.

To Thrustmaster, for keeping my feet busy while I was shooting cross-wind approaches.

A grand, special thanks to the people on the Flight Simulation Forum on CompuServe, for taking Flight Simulator into the information age.

And especially to Bruce Artwick, who made it all possible.

They all deserve heartfelt praise and gratitude.

Introduction

Who Is This Book For?

Whatever level of experience you have, this book can enhance your enjoyment of Microsoft Flight Simulator by providing helpful tips, hints, and practice sessions that are educational and entertaining. You can get the most from *Adventures in Flight Simulator* if you follow the path that best suits your experience:

If You've Never Used a Computer Before...

If you're new to computers, some of the concepts in this book may seem difficult. However, if you are capable of withdrawing money from the local ATM, you should be able to figure out which keys to press on a computer keyboard to get started. This book, along with the *Flight Simulator Pilot's Handbook*, and maybe some assistance from a kind friend with computer experience, will be enough to get you behind the controls of Flight Simulator. With this book's instruction, you will be flying gracefully over the World Trade Center in Chicago with enough skill to impress any other ATM operator.

With the *Flight Simulator Pilot's Handbook* open to Appendix D, "Keyboard Summary," begin reading this book from the very first chapter. Don't worry about making mistakes, and try the introductory flight a few times until you become comfortable with controlling the aircraft. Then you can move on to the next chapter. Follow the chapters sequentially. By the time you get to the Adventures, you should be very comfortable with using and flying Flight Simulator 5.

If This Is the First Time You Have Tried Flight Simulator...

It's never too late to start! Actually, there is nothing difficult about flying a plane, and Flight Simulator 5 makes it even easier.

Begin by reading and following the instructions in the first chapter to become familiar with the controls and concepts of flying with Flight Simulator. Chapters 2 through 5 will demonstrate the features of Flight Simulator 5 while you enjoy the various flight lessons. Then, as you become comfortable with flying, you can try your skills in the many Adventures in the second part of the book.

If You Can Take Off But Never Land...

This is a common dilemma for beginning Flight Simulator pilots. Fortunately, the new features of Flight Simulator 5 make it a lot easier to get into the air and enjoy the sights, and you don't have to worry about landing at all, because the aircraft will land itself if you tell it to.

By following the steps for the introductory flight in the first chapter, you can learn how easy it is to control the aircraft, fly around and look at the scenery, and watch as the aircraft lands itself. The second chapter will give you the instruction and practice necessary to perform a "squeaky clean" landing on your own. From there you can go on to whatever section of the book appeals to you, although you will get the most enjoyment from the book if you follow the chapters in sequence.

If You Can Fly Around, Look at the Sights, and Land, But Don't Understand the Instruments...

Hey! Did the Red Baron use instruments? Well, only a few. But then he didn't do a whole lot of night flying, either. When you have an understanding of the instruments, you can stop flying "the poor man's IFR" (I Follow Roads) and begin flying real IFR (instrument flight rules). With the realistic weather patterns that are now available in Flight Simulator 5, it is possible to take off on a clear day and end up in solid cloud cover before reaching your destination. Being able to use the instruments will keep you from getting lost, and can prevent crashes caused by an improper aircraft attitude.

Chapters 1 and 2 talk about the basic flight instruments and how to use them. Chapter 4 discusses and demonstrates the use of the rest of the instruments for navigation and precision control of the aircraft.

If You're a Professional Flight Simulator Pilot...

You might want to jump right into the "Engine Out over Innsbruck" flight in the Adventures section of this book and try the flight with mountains looming, an overcast sky blocking your view, and the engine dead at 9000 feet. You may find that you will even learn a little more about Flight Simulator and its new features by reviewing the lessons of the book and flying some of the flights in Chapters 4 and 5.

If You're a Real Pilot and You Think Flight Simulator Is a Toy...
Have you *really* taken a look at Flight Simulator 5? Flight Simulator 5 is the most authentic visual and instrument flight simulation package for microcomputers. With the use of third-party flight control systems (yoke and rudder pedals), sectional charts, and approach charts, you can simulate just about any flight situation possible. Just try out the last flight in Chapter 4, "Instrument Flight" and some of the IFR Adventures in the second part of the book—you'll find that Flight Simulator 5 can help you keep your flight and navigation skills sharp. Also, be sure to read the preface for an account of how Flight Simulator has helped me to become more comfortable with flying real aircraft.

Before You Begin
A Note About Memory
This book assumes that you are using the full memory required for running Microsoft Flight Simulator 5. Although Flight Simulator will run using less memory, its functions are limited and you may find that some of the features discussed in the book are not available to you. In addition, the colors described in the text and reproduced on the screen shots in this book may not match those on your screen if you are running Flight Simulator with limited memory. Refer to the Appendix at the back of this book or to the *Flight Simulator Pilot's Handbook* for information on how to configure your memory to optimize the performance of Flight Simulator 5.

The Pause Key
One of the most important features Flight Simulator offers is one that is not available in a real aircraft—the pause key. When you press P on your keyboard, Flight Simulator pauses in midflight, no matter where you are. This can be especially useful while you are reading this book. For example, you can use the pause key to suspend an example flight while you read and learn about the next topic. (After flying through a lesson or an adventure, you can always come back to the flight and go through it again without having to reread the entire lesson or description. Just follow the action steps.) If you are a beginner, you'll probably find yourself using this key more than any other feature until you get the hang of Flight Simulator.

Conventions Used in this Book

Before you begin flying with *Adventures in Flight Simulator*, you should become familiar with some of the text conventions used in this book:

User entry   Anything you are asked to type in, such as a longitude setting or the name of a situation you want to save, appears in boldface type.

Simultaneous keypresses   Keyboard keys that must be pressed at the same time to achieve a desired effect are shown linked with a hyphen. For example, pressing Shift-keypad 8 (that is, holding down the Shift key and pressing keypad 8 at the same time) will switch Spot view to the front of the airplane.

Using the mouse   Most mouse actions in this book require you to either click or drag. To click on an item using the mouse, move the mouse until the pointer is over the item you wish to select, and then click the left mouse button once. To drag an item using the mouse, move the mouse until the pointer is over the item you wish to drag, click and *hold down* the left mouse button, and move the mouse until the item or border has been dragged to where you want it.

Selecting   When the word *select* is used in this book, it usually applies to either a menu item or a dialog box selection. To select a menu item, either click on it with the mouse, or use the cursor keys to highlight it, and then press Enter. To select an item in a dialog box, click on its check box with the mouse, or use the cursor keys to highlight it, and then press Enter.

The Aircraft Controls

Flight Simulator 5 supports a variety of add-on products for flight control—joysticks, yokes, mice, trackballs, rudder pedals, and other products. It is impossible to discuss all of them in the context of this book. To account for all the possibilities, this book uses some generic terms to represent the various flight control add-on options. The generic terms and their descriptions follow:

The Yoke

Because the yoke of an aircraft is used for controlling the pitch (the up and down movement of the nose) and the roll (bank), the term "yoke" will be used to represent the physical controls. For example, the instruction "apply gentle left pressure to the yoke" will have the following meanings for the various devices:

Joystick Apply gentle left pressure on the joystick.

Keyboard Press the left cursor key on the numeric keypad or the cursor keypad once or twice.

Mouse Move the mouse gently to the left.

Yoke Turn the yoke gently to the left.

You'll probably need a few practice sessions with Flight Simulator to get the appropriate feel for your specific controls.

The Rudder

Although you won't use the rudder until later in this book, it is still important to understand how to control it. The instruction "apply right rudder pressure" will have the following meanings for the various devices:

Joystick (Not applicable unless a second joystick is attached and configured for use as a rudder. If you have only one joystick, see "Keyboard," below.) Apply right pressure to the second joystick.

Keyboard Press the keypad Enter key (on the right side of the numeric keypad).

Mouse Not usable as a rudder. See "Keyboard," above.

Rudder pedals Apply pressure to the right pedal.

The Throttle

You'll use the throttle frequently in the various flights in the book, so you should become comfortable with your configuration for throttle control. The instruction "reduce the throttle to 50-percent power" will have the following meanings for the various devices:

Second Joystick (Not applicable unless a second joystick is attached and configured to be used as a throttle.) Apply back pressure on the second joystick until the throttle knob on the aircraft instrument panel is in the middle position.

Keyboard If you have an enhanced keyboard (the function keys are on the top), press the F2 key repeatedly until the throttle knob on the aircraft instrument panel is in the middle position. If you have a standard keyboard (the function keys are on the left), press the F8 key repeatedly until the throttle knob is in the middle position. If you are using the numeric keypad, press the PgDn key repeatedly until the throttle knob is in the middle position.

Mouse Put the mouse cursor over the throttle knob on the aircraft instrument panel, hold down the left mouse button, and drag the throttle knob to the middle position.

Throttle on joystick Turn the throttle knob on the joystick toward you until the throttle knob on the aircraft instrument panel is in the middle position.

Throttle control on yoke Use the slider control or T-shaped push/pull knob to reduce the throttle until the throttle knob is in the middle position.

The Instrument Panel

Many of the instruments on the aircraft instrument panel can be controlled with the mouse pointer by either dragging (such as for the throttle, mixture, and propeller) or clicking (such as for the navigation radios, OBIs, etc.). For example, you can increase the radio frequency by clicking on the right side of the numeric display and decrease it by clicking on the left side of the numeric display. The same applies to the knobs for the OBI instruments: Clicking on the right of the knob will increase the heading setting, and clicking on the left of the knob will decrease the heading setting.

These same instruments can also be controlled with the keyboard. Appendix D in the *Flight Simulator Pilot's Handbook* contains the keyboard keystrokes for controlling Flight Simulator.

In the first couple of chapters, the necessary keystrokes are provided in the detailed steps for each flight.

Other Controls

Most of the other Flight Simulator controls are implemented with keyboard keystrokes. Some of the other controls that are needed for the flights are:

Elevator trim Keypad 1 (End) and keypad 7 (Home).

Flaps The keys you need to press to activate the flaps depend upon the style of your keyboard. Refer to Appendix D in the *Flight Simulator Pilot's Handbook* for a description of the keyboard functions.

Brakes The . (period) key. Ctrl - . (period) applies and releases the parking brakes.

Landing Gear The G key.

Carb Heat The H key.

Flight Settings

Before you begin any flight in this book, you will need to enter specific flight settings to configure the environment for the flight. These settings will include the following types of information:

Setting	Description and Menus
Aircraft	The type of aircraft that will be used for the flight. Menu: Options, Aircraft
Latitude and Longitude	The exact position of the aircraft in the simulated world. Accurate to real-world navigation settings. Menu: World, Set Exact Location
Altitude	The height of the aircraft above sea level. (The aircraft can be sitting on the ground but still be above sea level.) Menu: World, Set Exact Location
Heading	The magnetic directional heading that the aircraft is facing. Menu: World, Set Exact Location
Auto Coordination	When this feature is on, the rudder and ailerons are locked together for coordinated turns and banks. When Auto Coordination is off, the pilot must control the rudder separately from the ailerons. Menu: Sim, Auto Coordination
Engine	These settings determine the condition of the aircraft engine, such as On, Off, Throttle Control, etc. Menu: Sim, Engine and Fuel
Time and Season	The exact time of day and the season of the year. Menu: World, Set Time and Season
Weather	These settings determine the various weather conditions for locations within the simulated world. Menu: World, Weather
Propeller	These settings determine how the throttle affects the constant-speed propeller or allow you to set the plane for fixed-pitch propeller. Menu: Sim, Realism and Reliability

Saving and Loading Situations

Any situation within Flight Simulator can be saved as a file for reloading at a future time, which is quite helpful if you need to stop a flight before it is finished and then come back to it later. You can also save a situation after entering the flight settings for each flight, which allows you to start the flight all over again for practice, or in case you make a mistake. After going through all the adventures in the book, you can reload each situation, change some of the flight settings, and fly the adventure with different times and seasons, weather conditions, reliability factors, or aircraft.

To Load an Existing Situation

1. Open the Options menu.
2. Select Situations.
3. Select the situation you want to load.
4. Click OK with the mouse or press Enter.

To Create and Save a New Situation

1. Open the Options menu.
2. Select Save Situation.
3. Type in a title in the Situation Title box.
4. Type in a description in the Description box (optional).
5. Type in a name for the file in the Filename box (optional). (If you do not type a filename, Flight Simulator will use the first eight characters of the title you typed in the Situation Title Box, adding the .STN extension.)
6. Click OK with the mouse or press Enter.

Flying Clubs

Many clubs and groups meet to discuss flight simulations. Because of limited space, this introduction will discuss only two of the best ways to stay in touch with others interested in flight simulation.

CompuServe

One of the best places to meet other Flight Simulator pilots is in the FSFORUM section on the CompuServe online service. The FSFORUM (Flight Simulation Forum) provides an electronic meeting place where people can exchange messages and files and where members can conduct online conferences about Flight Simulator. Special topics of discussion include:

- **General aviation**
- **Air traffic control**
- **Scenery design**
- **Aircraft and adventure design**
- **Hardware**
- **Flight instruction**
- **Fly-ins**
- **Hangar talk**

CompuServe also has a Modem Lobby where people can meet and get together for multi-player flight using Flight Simulator.

Contact CompuServe for more information about membership.

MicroWINGS

MicroWINGS, also known as the International Association for Aerospace Simulations, is a great organization for keeping in touch with the flight simulation industry. MicroWINGS offers a regularly published magazine, a special section on CompuServe, a bulletin board system (BBS), discounts on flight simulation products, distribution of freeware/shareware software, a multi-player database, contests, drawings, and an annual conference.

It is worthwhile to join the organization for the conference alone. Many new products are announced and displayed at the conference each year, and it is a good place to meet the people behind the products. If you can't attend the conference, you can read about it, along with the many tips and reviews, in MicroWINGS Magazine.

For more information, contact:

>MicroWINGS
>381 Casa Linda Plaza #154
>Dallas, TX 75218
>(214) 324-1406

Chapter **1**

Introductory Flight

"Beautiful plane," the instructor remarks to the potential student. *"It's a real nice plane to fly, too. A Cessna Skylane R182. Would you like to go for an introductory flight?"*

"A ride?" the prospect asks, with the look of someone who has always dreamed of flying an airplane.

The instructor grins, "Nope, a flight! I'll take you up, give you the controls for awhile, and let you see whether you like it or not."

"Well...sure! I'd love to go up."

"Great! Just wait here for a minute while I get the keys, and then we'll head on out."

The Preflight

Unlike traveling in your car, where you can just get in, start the engine, and go, traveling in an aircraft takes a little preparation. For starters, we need to have a flight plan. Normally, a flight plan is a schedule of the planned flight activity, which the pilot files with the local flight service station. In this book, a flight plan will simply describe a prospective flight and list the activities it will involve.

Flight Plan: For this first flight, we will depart from Meigs Field in Chicago in a Cessna R182. The flight will be local and will introduce you to climbs, straight and level flight, turns and banks, and visual navigation. We'll head north over the western coast of Lake Michigan and will then turn inland over downtown Chicago for a final automatic landing approach into Meigs Field.

✈︎✈︎✈︎

The next step in preparing for a flight is to perform a preflight inspection of the aircraft. To do this in Flight Simulator, you first need to enter the flight settings. Follow the instructions on page 4 to enter the settings shown in the table.

Flight Settings

Location	Meigs Field, on the ramp
Aircraft	Cessna Skylane RG R182
North/South Latitude	N041°51'37.1996
East/West Longitude	W087°36'33.3996
Altitude	594 feet
Heading	91°
Auto Coordination	On
Engine	Off
Season	Spring
Time	9:30 a.m.

Aeronautically Speaking: The flight plan helps air traffic control keep track of the aircraft during its flight. It also serves as a safety measure, because it includes the estimated time of arrival (ETA) at the destination airport. If the aircraft does not arrive within a designated period of the ETA, various services, such as the Civil Air Patrol, conduct a search along the planned flight path. Thus, it is important for the pilot to "close" the flight plan on arrival at the destination in order to prevent an unnecessary search for the aircraft.

Chapter 1: Introductory Flight

Wing Tip:
Before you begin any flight, be sure to deactivate Slew, if it is on. Open the World menu and look to see if there is a check mark next to the Slew option. If there is, click on the option to deselect it.

Enter the Flight Settings

1. Open the Options menu, and select Aircraft.
2. Select the Cessna Skylane RG R182, and click OK.
3. Open the World menu, and select Set Exact Location.
4. In the North/South Lat. box, type **N041*51'37.1996**, and press Enter. (Because most keyboards do not have a degree symbol, Flight Simulator accepts an asterisk to signify degrees in longitude and latitude settings.)
5. In the East/West Lon. box, type **W087*36'33.3996**, and press Enter.
6. In the Altitude box, type **594**, and press Enter.
7. In the Heading box, type **91**, and press Enter.
8. Click OK.
9. Open the Sim menu, and be sure that there is a check mark to the left of Auto Coordination. If not, click on that selection or press A.
10. Open the Sim menu again, and select Engine and Fuel.
11. Under Engine 1, click in the box that says Both and change it to display Off.
12. Click OK.
13. Open the World menu, and select Set Time and Season.
14. Set Season to Spring.
15. Select Time of Day, and set Time of Day to Day.
16. Select Set Exact Time. In the Hours box, type **9**, and press Enter. In the Minutes box, type **30**, and press Enter.
17. Click OK.

The Walk-Around

Now let's check out the aircraft. A methodical, orderly inspection of the airplane is one of the most important parts of the preflight procedures. Ordinarily, a pilot walks around the plane and follows a detailed checklist. To do this in Flight Simulator, we must change to a full external view of the plane.

Set Spot View

1. Open the Views menu, and select View Options.
2. Under View 1, change Cockpit to Spot. In the Zoom box, type **1**, and press Enter.
3. Click OK.
4. Open the Views menu, and select Set Spot Plane. In the Distance box, type **55**, and press Enter. In the Altitude box, type **0**, and press Enter.
5. Use the mouse or the arrow keys to move the dot in the View Direction box so that it is directly behind the plane.
6. Click OK.

Now that the view has been configured for our walk-around, let's start with the cockpit. Press S to see inside.

On the right side of the instrument panel, check to be sure that the lights and strobes are on. If they are not on, click on their respective on/off indicators to turn them on. Now look at the engine instruments. (If you are in VGA mode, you will have to press Tab to see them all.) Make sure that the *magnetos* (explained on page 14 in an Aeronautically Speaking note) are off, and check the fuel levels of both the left and right tanks to be sure that they are full. Press the W key to maximize your active window.

Now we need to walk around the aircraft to make sure everything looks okay. Press S twice to move to the back of the aircraft.

Now lower the flaps to the full extended position by pressing the F8 key.

Notice that the flaps have become extended in the full down position. We will use the flaps later when we come in for a landing. They will help to slow the plane down while creating extra lift to keep us flying.

Wing Tip:
The keys you press to activate the flaps vary depending on whether your keyboard is standard (with the function keys at the side) or enhanced (with the function keys at the top). This book assumes you have the extended keyboard. See Appendix D in the Flight Simulator Pilot's Handbook for a description of standard keyboard key functions.

Chapter 1: Introductory Flight

Spot views of the Cessna during the walk-around.

Lower the Flaps and Start the Walk-Around

1. Press S to switch to Cockpit view.
2. Check the lights and strobes on the instrument panel to be sure they are on. If they are not on, click on their respective on/off indicators to turn them on.
3. Check fuel levels on the fuel gauges.
4. Check to be sure the magnetos are off. If they are not off, click on the Mags Off indicator to turn them off.
5. Press S twice to return to Spot view, and press W to maximize the active window.
6. Press F8 to lower the flaps.

Here, at the tail of the aircraft, you will notice two different control surfaces. One is the rudder, and the other is the elevator. The rudder is attached to the vertical stabilizer. The rudder controls the *yaw* of the airplane, or the movement of the nose to the left or right. The elevator, which is attached to the horizontal stabilizer, controls the *pitch* of the aircraft—that is, the movement of the nose up or down. The use of these controls changes the *attitude* of the aircraft. *Attitude* refers to the position of the aircraft in the air, relative to the ground. For example, if we were to push the elevator down while flying so that the pitch of the nose was pointing below the horizon, we would say that the aircraft was in a "nose-down attitude."

During the introductory flight, you'll see how the control surfaces affect the attitude of the aircraft.

With a real aircraft, we would be checking at this point to be sure that the rudder and elevator could move freely and that they were properly attached to the aircraft. But because the programmers of Flight Simulator have assured me that the control surfaces will not fail during flight, we can proceed to the right side of the aircraft.

Change the View to the Right Side of the Aircraft

1. Open the Views menu, and select Set Spot Plane.
2. Use the mouse to move the dot in the View Direction box to the right side of the aircraft.
3. Click OK.

From here, we can see the green navigation light, the ailerons, and the flaps (which are still extended). The ailerons are the control surfaces next to the flaps, on the outward trailing edge of the wing. The ailerons control the *bank*, or roll, of the aircraft. When the aileron on the right side of the aircraft is up, the aileron on the left side is pointed down. This difference causes the aircraft to bank to the right. In conjunction with the rudder, these controls allow the aircraft to turn. We will talk more about turns when we get into the air. Again, with a real aircraft we would also be checking the tires and the brakes, draining the fuel to check for water, and making sure that the control surfaces could move freely and were attached correctly to the aircraft.

Let's move around to the front of the aircraft.

Change the View to the Front of the Aircraft

1. Open the Views menu, and select Set Spot Plane.
2. Move the dot in the View Direction box to the front of the aircraft.
3. Click OK.

Wing Tip:
You can use the keypad keys, with the Shift key, as a short-cut to change the direction of your Spot views. Shift-keypad 6 provides a view of the right side of the aircraft, Shift-keypad 4 provides a view of the left side of the aircraft, etc.

Chapter 1: Introductory Flight

On a real aircraft, we would be checking a lot of things at this point. We would check the engine oil level, look for nicks or cracks on the propeller, look for obstructions in the engine compartment, check for loose wires or parts, and drain any water from the fuel tanks. But we don't need to do this in Flight Simulator, so let's just retract the flaps by pressing the F5 key and go on to the left side of the aircraft.

Change the View to the Left Side of the Aircraft

1. Press F5 to raise the flaps.
2. Open the Views menu, and select Set Spot Plane.
3. Move the dot in the View Direction box to the left side of the aircraft.
4. Click OK.

About all we can do here is to notice that the navigation light is red. On the right side of the aircraft, the light is green. This helps other pilots see the aircraft at night and tells them which way the plane is flying. For example, if we were flying at night and saw a red rotating beacon and a green light, we would know that it was an airplane and that it was traveling from left to right. We would not be able to see the solid red navigation light on the other side of the airplane (assuming that the plane was flying level and not turning).

Now that we've finished the external preflight inspection, it's time to climb in and continue working through our preflight checklist for starting the aircraft. Press S to set the view to inside the cockpit, and press W to minimize the active window so we can see the instrument panel.

Scenery

In the Cockpit

With the instrument panel before us, let's first make sure that the parking brakes are set. Press Ctrl- . (period) to turn on the parking brakes.

We should also check the fuel selector valve to be sure that it is set to both tanks. This ensures that the engine will continue to get fuel if one of the tanks runs dry or becomes stopped up, and it also keeps the plane balanced with the same amount of fuel in both wings.

Set the Brakes and Check the Fuel Selector

1. Press S to return to Cockpit view, and press W to minimize the active window.
2. Press Ctrl- . (period) to apply the parking brakes if they are not already on.
3. Open the Sim menu, and select Engine and Fuel.
4. Check the box next to Fuel Selector to be sure that it displays All.
5. Click OK to apply these settings.

Now we need to be sure that the carburetor heat is set to off. The carburetor heat prevents the carburetor from icing up when conditions are cold or humid or both. The control for it is located just above the ignition switch on the engine panel.

Turn Off the Carburetor Heat

1. Locate the Carburetor Heat control on the instrument panel.
2. If the Carb indicator is not already off, click on it or press H to turn it off.

Wing Tip:
When changing radio frequencies with the mouse, click on the left side of the number display to decrease the frequency, and click on the right side of the display to increase the frequency.

Aeronautically Speaking:
The weather information broadcast by ATIS is updated regularly. Each new weather update is assigned an alphabetic ID, such as Alpha, Bravo, Charlie, and so on. A reference to this ID—"information Alpha," for example—lets the controller know whether the pilot is using the latest weather information.

Okay, we're just about ready to start the engine. Move the throttle up about one-quarter inch (a few notches) by pressing keypad 9 five times. Then turn the ignition switch to Start.

Start the Engine

1. Press keypad 9 five times to move the throttle up about one-quarter inch.
2. Click on Start in the Mags indicator box or press M and the + (plus) key at the top of the keyboard four times.
3. Use keypad 3 and keypad 9 to adjust the throttle until the RPM indicator reads just over 1000.

Congratulations, you have just successfully started the engine! It is important to keep an eye on the oil temperature and oil pressure gauges just after you start the engine to make sure that they come up to normal readings in the green area. In real-world aviation, if the engine gauges do not indicate green, it is usually a sign of engine problems, and it is best to turn off the engine right away.

How are you feeling so far? A little nervous? That's okay. Once we get into the air and you find out how easy it really is, your palms will stop sweating. Maybe!

Taxi Time

Before we taxi to the runup area, we need to check ATIS (Automatic Terminal Information Service) for the latest weather and airport information and then contact ground control to get permission to taxi. So first, set the frequency for ATIS. (If you're in EGA or VGA display mode, you'll have to toggle back to the radio panel by pressing Tab.)

Set the COM Radio Frequency to ATIS

1. Open the Nav/Com menu and select Communication Radio.
2. In the COM 1 Frequency box, type **121.30**, and press Enter.
3. Click OK.

Be sure to pay attention to the scrolling transmission from the tower. Now I'll call the tower to let them know we are ready.

"Meigs Ground, this is Cessna N22287, with information Alpha, ready to taxi to runup for a departure on runway 36."

"Cessna 287, you are cleared to taxi to the runup area. Please notify ground when your runup is complete."

"Cessna 287."

Wing Tip:
You can use the W key to toggle between a maximized and minimized active window any time during a flight.

It is a lot easier to taxi in an external view, so let's switch to one that is a little more manageable.

Set Spot Plane View and Change to Spot View

1. Open the Views menu, and select Set Spot Plane.
2. In the Distance box, type **100**, and press Enter.
3. In the Altitude box, type **25**, and press Enter.
4. Move the position dot to directly behind the aircraft.
5. Click OK to apply these settings.
6. Press S twice to switch to Spot view.

Now we're ready to taxi out to the runup area. If at any time you get confused or frustrated, press P to pause and take a break. Or just taxi around the airport and get comfortable with the throttle and the controls. You can always come back to this spot and continue from here after you become more familiar with the aircraft.

Go ahead and release the parking brakes and give the aircraft enough throttle to start moving. Once the plane is moving, back the throttle off a little to keep from going too fast. Taxi up to the taxiway, and then make a right turn. The taxiway is the beige strip of pavement directly ahead of us. The gray and white strip beyond

Chapter 1: Introductory Flight

Wing Tip:
This would be a good time to save the situation. Then you can always come back to this point and continue with the lesson. For instructions on how to do this, refer to "Saving and Loading Situations" in the Introduction.

the taxiway is the runway. Be sure to watch for any other aircraft or vehicle traffic on the taxiway. Before we reach the end of the taxiway, make a turn to the right, proceed another 50 feet, and then make a U-turn to face toward the taxiway. Use the brakes to stop here.

Taxi to the Runup Area

1. Press the . (period) key to turn off the parking brakes.
2. Apply enough throttle to get moving by pressing keypad 9 several times.
3. Once you are moving, bring the throttle back down to 1000 RPM by pressing keypad 3 several times.
4. Taxi up to the taxiway, and then make a right turn by moving the yoke to the right.
5. Use the brakes by pressing the . (period) key to slow down or stop if necessary.
6. Before you reach the last left turn to the runway, make a right turn.
7. Proceed a short distance into the runup area, and then make a U-turn to face toward the taxiway.
8. Press the . (period) key to apply the brakes and come to a stop.
9. Press S to switch back to Cockpit view.

The Runup

This is the runup area. This is where we run some tests on the aircraft to make sure that the engine and the controls are performing correctly before we actually take off. Usually we would face the aircraft into the wind during the runup, but I like to face just a little to one side, which lets me see whether any planes are approaching to land and also allows me to taxi straight up to the hold line of the taxiway.

For the runup, we will first need to check the flight controls to make sure they are functioning correctly. To do this, move the yoke as far as possible in all directions and then back to the neutral position. Watch the control position indicators on the instrument panel to see the movement of the controls. Make sure that the indicators are moving to their fullest extent.

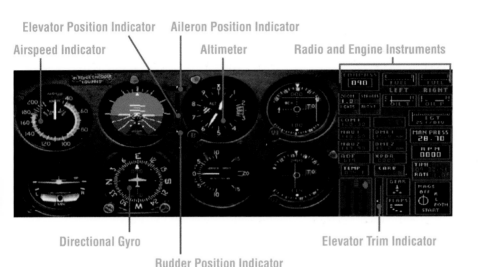

Aeronautically Speaking: Normally, during the runup procedure with a real aircraft, the pilot checks visually to be sure that the control surfaces are actually moving in conjunction with the controls. One thing the pilot would watch for would be an incorrect movement of a surface, such as the rudder moving right when the left rudder is pushed. Such movement would indicate that the internal control cabling and mechanisms were hooked up incorrectly during maintenance or repairs. The pilot also checks to be sure that the control surfaces are moving to the fullest possible extent, thus demonstrating that nothing is restricting the free movement of the controls. A safe pilot checks the controls before every flight.

Test the Movement of the Controls

1. Move the yoke to the left as far as it will go while you watch the control indicators. Continue to watch the indicators as you complete the test.
2. Keep the yoke left while moving it all the way forward.
3. While the yoke is all the way forward, move it all the way to the right.
4. While the yoke is all the way to the right, move it as far back as possible.
5. Return the yoke to the normal neutral position.

Now that we have felt the movement of the controls and have observed their effect on the indicators, let's see how the controls affect the flight of the aircraft. We have already discussed the effects that the rudder, ailerons, and elevator have on the aircraft, so I will be brief in explaining how to control these surfaces. The control yoke controls the movement of the ailerons and elevator. Pulling back on the yoke causes the elevator to move upward, thus causing the nose of the airplane to pitch

Chapter 1: Introductory Flight

Aeronautically Speaking:
The magnetos supply electric current to the spark plugs, which ignite the air/fuel mixture in the engine. For safety reasons, an airplane normally has two magnetos. If one of the magnetos fails during a flight, the other magneto can be relied upon to get the airplane to safety. Each cylinder of the aircraft engine has two spark plugs for the same safety reasons. Each magneto supplies current to one of the two spark plugs. During the magneto test, the RPM drops because only one of the spark plugs ignites the air/fuel mixture. Such a drop in the RPM during a flight could indicate that one of the magnetos has failed.

up. Pushing forward on the yoke lowers the elevator, which makes the nose of the airplane pitch down (unless you are flying upside down, but I'll save that for another flight). Moving the yoke to the left moves the left aileron up and the right aileron down, causing the plane to roll to the left. Moving the yoke to the right has the opposite effect.

Pushing on the left rudder pedal moves the rudder to the left, causing the airplane to yaw to the left. Pushing on the right rudder pedal has the opposite effect. On this flight we will be using Auto Coordination, which "locks" the rudder and ailerons together, providing coordinated turns. When you bank for a turn, the appropriate amount of rudder will automatically be applied. Auto Coordination makes flying a lot easier for new pilots because it is a little difficult to determine, without experience, how much rudder to apply for a given amount of bank.

The next runup tests we need to perform are the magneto and carburetor heat checks. To do this, we'll need to apply the parking brakes and then increase the throttle to 1700 RPM. Notice that the nose of the airplane will pitch down a little from the thrust of the propeller, but don't worry about the plane going anywhere; the parking brakes will keep it in place. Now we need to move the ignition switch to the left magneto. Notice that the RPM (revolutions per minute) drops slightly. The drop should be no more than 130 RPM. Move the switch back to B for both magnetos. The RPM should come back up to the original setting of 1700. Now turn the ignition switch to the right magneto. You should notice the same type of drop in the RPM. Move the switch back to the B position, and leave it there.

Test the Magnetos

1. Apply the parking brakes by pressing Ctrl- . (period).
2. Adjust the throttle with keypad 9 until the RPM reads just over 1700.
3. If you are in VGA mode, press Tab to display the engine instruments.
4. Click on the L in the Mags indicator box or press M and the – (minus) key at the top of the keyboard to select the left magneto, and watch for the drop in the RPM.
5. Click on Both in the Mags indicator box or press the + (plus) key at the top of the keyboard to select both magnetos, and watch the RPM come back to the previous setting.
6. Click on the R in the Mags indicator box or press the – (minus) key at the top of the keyboard twice to select the right magneto, and watch for the drop in RPM.
7. Click on Both in the Mags indicator box or press the + (plus) key at the top of the keyboard twice to select both magnetos. The RPM will return to the previous setting.

After we're sure that the magnetos are reliable, we can test the carburetor heat. We'll do this by keeping the throttle at 1700 RPM and then pulling the carburetor heat switch. Again, you will notice a drop in the RPM. Turn the carburetor heat off, and the RPM will return to the original setting.

Test the Carb Heat

1. Click on the Carb on/off indicator or press H to turn on the carburetor heat, and notice the drop in the RPM.
2. Click again on the Carb on/off indicator to turn off the carburetor heat, and watch the RPM return to the previous setting.
3. Press keypad 3 several times to reduce the throttle to 1000 RPM.
4. If you are in VGA mode, press Tab to return to the communications instruments.

Aeronautically Speaking: The carburetor heat system of the aircraft provides warm air from the exhaust system to the carburetor. Because air is compressed as it enters the carburetor, it becomes very cold. If the air is humid, ice can form around the carburetor intake and block the flow of air. This can occur even if the air temperature outside is warm. The carburetor heat system, when on, keeps the air intake of the carburetor warm enough to prevent ice from forming.

Chapter 1: Introductory Flight

Departure

Our runup is now complete, and we are ready for departure. We just need to contact the tower to let them know that we are ready.

"Meigs Ground, Cessna 287 with information Alpha, runup complete."

"Cessna 287, be alert to any traffic on the taxiway. Maintain right of way and taxi to and hold short of runway 36."

"287 holding short."

Before you taxi up to the runway, look to the left to make sure that no other aircraft are also taxiing for departure. When you are sure that the taxiway is clear, go ahead and taxi up to the hold line (the line on the taxiway just short of the runway).

Watch for Other Aircraft and Taxi to the Runway

1. Hold down the Shift key and press keypad 4 for a view out the left side of the aircraft.
2. Hold down the Shift key and press keypad 7 for a view out the left front of the aircraft.
3. If the taxiway is clear, hold down the Shift key and press keypad 8 to return to the front view.
4. Press S twice to switch to Spot view.
5. Release the parking brakes by pressing the . (period) key.
6. Increase the throttle by pressing keypad 9 until the airplane starts to roll forward; use keypad 3 to decrease the throttle if necessary.
7. Use the yoke to steer the airplane to the hold line, just short of the runway.
8. Using keypad 3, reduce the throttle to about 1000 RPM; press the . (period) key to apply the brakes for stopping.

Maintain the throttle at about 1000 RPM, and keep your feet on the brakes by holding down the . (period) key. Keep an eye out for any aircraft coming in for a landing.

"Cessna 287, you are cleared for departure."

"287, thank you."

Okay, we've got the go-ahead for takeoff. Release the brakes, throttle up just enough to get the plane moving, and taxi onto the runway. Line up on the runway so that the nose of the airplane is pointed down the center of the runway.

Taxi onto the Runway

1. Press keypad 9 to increase the throttle just enough to get the plane moving.
2. Taxi onto the runway, and then turn left to line up on the center of the numbers. (You should be pointing down the center.)

Here's the big moment. Go ahead and give it full throttle, keep the nose centered down the runway, and then gently pull back on the yoke at 70 knots. After the plane lifts off the runway, raise the landing gear.

Take Off

1. Press S to switch to Cockpit view.
2. Hold down keypad 9 until the throttle is in the full up position.
3. Gently use the yoke to keep the plane pointed down the center of the runway.
4. At 70 knots, gently pull back on the yoke until the horizon is just barely above the instrument panel.
5. After liftoff, raise the gear by pressing G.

We Have Liftoff!

The view of Meigs Field shortly after departure.

All right! You just lifted off the runway. That wasn't so hard, was it? Remember to keep the nose pitched so that the horizon line of the ground appears just above the instrument panel. This should give us a climb speed of about 80 knots. As we climb, let's get a good look at the airport from the air.

View the Airport

1. Press S twice to switch to Spot view.
2. Hold down the Shift key, and press keypad 8.
3. Press S to return to Cockpit view.

Now, that's what I call a great view of Meigs Field. It's pleasant to look out the windows, but it's easy to lose yourself in the scenery and then find out that the airplane has altered its original attitude. So you must always check your references to make sure that the aircraft is in the correct attitude. Check to be sure that the horizon is still sitting just a little above the instrument panel. Go ahead and look through the various windows and views now, but do check periodically from the

front view to be sure that the airplane is in the desired attitude. Remember that you can use W to toggle between a maximized or minimized active window during flight.

Look at the Various Views

1. Hold down the Shift key and press any of the keys on the numeric keypad to get various views from inside the cockpit. (Shift-keypad 5 gives you a view of what is directly below you.)
2. Press S twice to switch to Spot view.
3. Hold down the Shift key and press any of the keys on the numeric keypad to get the various external views.
4. Press S once to return to Cockpit view.

Feel free to use the pause key (P) so that you can sit back and relax for a few minutes. While the simulation is paused you can still change the views (only in Cockpit view), but you won't have to worry about the attitude of the aircraft. Then, when you are ready to resume flying, just press P again.

The altimeter.

The throttle controls.

About this time we should be approaching 2000 feet in altitude, which should be reflected by the altimeter, shown at left.

After the small hand of the altimeter reaches 2 and the large hand passes over the 0 (for 2000 feet), reduce the throttle until the throttle control is about one-fifth of the distance from the top. You will notice that the nose of the airplane begins to drop. It's okay! The plane is just starting to level off because you have reduced the power. When the altimeter shows 2400 feet, drop the nose of the airplane gently. When the altimeter shows that you are not gaining or losing any altitude, look at the position of the horizon in relation to the instrument panel. Remember how this looks, because it provides a visual reference for straight and level flight.

Level Off

1. Reduce the throttle by pressing keypad 3 until the throttle knob is one-fifth of the distance from the top.
2. When the altimeter reads 2400 feet, drop the nose of the plane by pushing gently forward on the yoke.
3. Try to keep the plane level at 2500 feet by pushing or pulling on the yoke as needed.

Depending on our current altitude, raise or lower the nose of the plane until we are flying at 2500 feet. Now try to hold this altitude and notice how it might require you to maintain constant pressure on the yoke in order to keep the plane level. After awhile, keeping the plane level can become quite tiring, a problem that is solved by the elevator trim control. This control keeps the aircraft trimmed to the desired attitude, eliminating the need to apply constant pressure on the yoke. Use keypad 7 and keypad 1 to adjust the elevator trim: Keypad 7 trims the nose of the airplane down, and keypad 1 trims the nose up.

Using the elevator trim, keep the aircraft flying level. If you need to use the yoke to compensate a little, go ahead. The main purpose of the elevator trim is not to force the aircraft to maintain a constant altitude but to make the job of keeping the plane level a little easier for the pilot. For the time being, don't worry about maintaining a constant altitude; just try to keep the plane as level as possible.

Trim the Aircraft for Level Flight

1. If needed, press keypad 7 to lower the nose of the aircraft.
2. If needed, press keypad 1 to raise the nose of the aircraft.
3. Compensate as needed with the yoke.

You might have noticed that as you dropped the nose, the airspeed of the plane increased. When the airspeed increases, the flow of air over the wing is greater, which provides more lift and thus causes the plane to climb again. Throughout the process of leveling off, you will need to continue making adjustments to the elevator trim and the yoke until the aircraft has stabilized in level flight. You'll notice also that the airspeed of the aircraft has increased to around 140 knots.

Another helpful hint for visually checking that the attitude of the aircraft is level is to look at the bottom of one of the wings. Hold down the Shift key and press the keypad 4 key to change your view to the left wing, and I'll show you what I mean.

We are now looking at the bottom of the left wing of the aircraft. You'll notice that the bottom of the wing is flat. Now notice how the horizontal line of the horizon looks in relation to the bottom of the wing. If the two horizontal lines are parallel, the pitch of the aircraft is level. If the spacing between the horizon and the bottom of the wing is greater in the front of the wing than in the back, the aircraft is currently in a nose-up attitude. If the spacing is greater in the back of the wing, the aircraft is in a nose-down attitude.

If the bottom of the wing is level with the horizon, it serves as a visual check that the pitch attitude of the aircraft is level. However, it is possible to be slightly banked even though the bottom of the wing seems level with the horizon. To ensure that the aircraft truly is level, you also need to check from the right side of the aircraft. Look out the right window, and notice the distance between the bottom of the

Aeronautically Speaking: Having the nose of the aircraft pointed up does not necessarily mean that the aircraft is climbing. It is a regular procedure in some types of landing approaches to have the nose of the aircraft pointed up over the horizon even as the aircraft is descending. This will be demonstrated in a future flight in which the aircraft is in a slow-flight attitude. But in the meantime, don't make the deadly assumption that you are climbing just because the nose is pointed toward the sky.

wing and the horizon. Then look out the left window again, and check to see if the distance is the same. If it is, the aircraft is in straight and level flight. If the distance on the left is smaller than the distance on the right, in which direction do you think the aircraft is banked? If you said, "to the left," you are correct. Actually, I'm just testing you to see whether you should stick to flying paper airplanes!

Visually Check for Straight and Level Flight

1. Hold down the Shift key and press keypad 4 to view the left wing.
2. Notice the distance between the bottom of the left wing and the horizon.
3. Hold down the Shift key and press keypad 6 to view the right wing.
4. Notice the distance between the bottom of the right wing and the horizon. If the distances are the same, and if the horizontal lines of the wings and the horizon are parallel, you are in straight and level flight.
5. Hold down the Shift key and press keypad 8 to return to the forward view.

Turn and Bank

Starting to get a little bored with flying straight and level? Ready for more of a challenge? Good! Because now it's time to try a turn and bank.

Before you begin with turns, let's see what happens during the turn itself. When the aircraft is banked, it will want to descend because the vertical lift has been reduced. You will still have lift, but the aircraft will be pointed in a direction other than up. So to make a turn, you will not only need to apply left pressure to the yoke, you will need to apply a little back pressure as well. With this in mind, go ahead and bank the aircraft to the left.

Start the turn by gently applying left pressure to the yoke. As you notice the nose of the aircraft begin to drop, pull back on the yoke a little to keep the nose in a level attitude. With the correct angle of bank, the horizon should touch the top left corner of the view screen. When you have reached this angle, center (neutralize) the yoke. The aircraft will gradually tend to return to level flight, so you will need to apply a little more pressure on the yoke to maintain the bank angle that you need for the turn. Keeping an eye on the vertical airspeed indicator or the altimeter will help you determine whether the aircraft is losing altitude during the turn. When the nose of the aircraft is close to pointing directly at downtown Chicago, gently apply right pressure to the yoke until the aircraft has leveled off.

Because you had to apply a little back pressure on the yoke during the turn, the aircraft will try to climb as soon as the wings are level and all the lift is pointed directly up again. You can compensate by applying a little forward pressure on the yoke.

Turn Left and Point Toward Chicago

1. Apply left pressure to the yoke and bank the aircraft until the horizon is touching the upper left corner of the view screen.
2. Gently apply a little back pressure on the yoke to keep the aircraft at the same altitude.
3. When the aircraft is pointed toward downtown Chicago, level off by applying right pressure to the yoke.
4. Use keypad 1 and keypad 7 to trim the aircraft for straight and level flight.

You might have ended the turn with the plane pointing off to the left of Chicago. If this happened, go ahead and turn a little to the right until you are pointed toward the center of the city. Turning too far is a common mistake. A new pilot often waits until the aircraft is pointed directly at the desired direction before starting to come out of the turn. The correct way is to start coming out of the turn a little before you actually reach the desired direction. Then, by the time the aircraft becomes level, it will be pointed in the right direction. This takes a little getting used to, but it will come with practice.

Over Chicago

Cruising over Chicago.

As we continue in straight and level flight, we can take some time to enjoy the sights—at least until we start to see the last of the Chicago high-rise buildings pass under the aircraft. Using what you have learned so far about the views, go ahead and examine the different views over Chicago.

After passing over the last high-rise building on the south side of Chicago, make a slight right turn until the aircraft is pointed toward the river. We'll be able to see that the river makes a right turn to the east. Don't worry about following the river, just stay pointed in the direction in which you are headed now. The directional gyro (compass) should be indicating a heading of somewhere around 190°, or just a little to the right of south. When the aircraft is straight and level again, look out the left rear of the aircraft toward Meigs Field.

Turn Toward the River Southeast of Chicago

1. After passing over the last visible high rise, gently bank to the right until the aircraft is pointed toward the river.
2. Level out, and check to see that your heading is around 190°.
3. Hold down the Shift key and press keypad 1 to change the view to see Meigs Field.

Keep flying straight and level until Meigs Field is in the center of the view screen. Remember, we are looking out the left rear of the aircraft. As soon as the south end of the runway at Meigs Field passes the center of the view screen, flip back to the forward view, and then make a left turn toward Lake Michigan. When the horizon is touching the upper left corner of the view screen, neutralize the pressure on the yoke (center the yoke), but use back pressure as necessary to keep the nose level. Start coming out of the turn at 100° by gently applying right pressure to the yoke. You should be level and finished with the turn when the directional gyro indicates 90°, due east. If you look out the left window of the aircraft, you will be able to see downtown Chicago and then Meigs Field off to the east. Reduce the throttle to 30-percent power. The nose of the aircraft will pitch down, but that's okay. This is called the descent, or *approach*, into the airport.

Start the Approach to Meigs Field

1. Keep flying straight and level, and watch for the south end of the runway at Meigs Field to pass by the center of the view screen.
2. Make a left turn by gently applying left pressure to the yoke until the horizon is touching the upper left corner of the view screen.

3. Remember to apply a little back pressure to the yoke to keep the nose level.
4. When the directional gyro shows 100°, begin banking to the right by applying right pressure to the yoke until the horizon is level.
5. The directional gyro should show 90° when the turn is finished and the wings are level with the horizon.
6. Look left by holding down the Shift key and pressing keypad 4, and notice the position of Meigs Field.
7. Switch back to the forward view by holding down the Shift key and pressing keypad 8.
8. Reduce the throttle by repeatedly pressing keypad 3 until the manifold pressure is around 12.

All right! You have successfully returned us to the airport. All you need to do now is turn over control of the aircraft by turning on the Land Me feature. Landing a plane takes a bit of practice, and it's a little much to expect of someone who is just starting out. But a good way to start learning about landing is to watch what is going on while someone else (or something else, in this case) does it. So press X to turn on Land Me, and watch—in various internal and external views—as the aircraft lands at Meigs Field.

Land the Aircraft
1. Press X to activate the Land Me feature.
2. Watch as the Land Me feature lands the aircraft.

What do you think of Flight Simulator 5 so far? And this is only the beginning! During this flight we covered the following topics:

Preflight inspection of the aircraft
The instrument panel
Communication with the tower
The runup
The takeoff
Climbs
Straight and level flight
Turns and banks
Visual navigation

Chapter **2**

Pilot Training and Private Lessons

The instructor points out the left window of the aircraft. "See how the wing of the plane stays pointed at the top of the building? Now look back at the turn coordinator and notice what the angle of the bank is. Take a quick peek and then watch the position of the wing over the building to make sure that you are maintaining that angle as you continue your turn around the building."

"Okay," the instructor directs, "now add a slight amount of right pressure to the yoke. What happens to the position of the building?"

"It starts to move behind the wing," the student replies.

"Correct! Now return to the original angle of bank, and then add a slight amount of left pressure to the yoke. Now what happens to the position of the building?"

"It moves forward of the wing."

"Good," praises the instructor. "What we are doing is called making a turn around a point. Practicing these will help you learn to control an aircraft precisely by using visual references. You're doing well, so let's try it again. See the Eiffel Tower up ahead? Let's do a turn around that point*!"*

Want To Be a Pilot?

Flying over the Eiffel Tower does sound like a lot of fun, doesn't it? And it's even more fun when you're doing the flying instead of paying someone to do it for you. As you discovered during the introductory flight in Chapter 1, it's not hard to fly an airplane. But to become a licensed pilot, you have to be familiar with the procedures for precision flying and navigation. Private pilot lessons can provide this type of training. In this chapter, we'll go on some training flights that will provide you with some of the skills you need to be a private pilot.

So if you're ready to begin training for your Flight Simulator private pilot certificate, let's get started.

Flight Plan: For this flight, we'll fly the Cessna R182 over Paris. This flying lesson will cover simple maneuvers, such as takeoff, turns around a point, standard two-minute turns, turns to a heading, climbs, and use of basic flight instruments. We'll use ground reference points while over Paris, and we'll conclude by buzzing the Eiffel Tower and the city buildings.

✝ ✝ ✝

Before we depart, enter the settings shown in the table. If you need help, follow the instructions below.

Flight Settings

Location	Paris
Aircraft	Cessna Skylane RG R182
North/South Latitude	N048°43'14.8065
East/West Longitude	E002°21'37.8710
Altitude	295 feet
Heading	78°
Auto Coordination	On
Season	Spring
Time of Day	Day
Weather—Clouds	
Base	10,000 feet
Tops	11,000 feet
Coverage	3/8, Scattered
Realism and Reliability	
Prop Advance	Fixed Pitch

Aeronautically Speaking: Currently, an applicant for a private pilot certificate must have at least 20 hours of flight instruction from a certified flight instructor and 20 hours of solo flight. The solo flight time includes 10 hours of cross-country flying, in which one of the flights must be longer than 300 nautical miles and have landings at a minimum of three airports.

Currently, the FAA does not recognize Microsoft Flight Simulator as an authorized training device for real-world flight. However, with the increased realism and functionality of version 5, Flight Simulator can still provide invaluable experience.

Enter the Flight Settings

1. Open the Options menu, and select Aircraft.
2. Select the Cessna Skylane RG R182, and click OK.
3. Open the World menu, and select Set Exact Location.
4. In the North/South Lat. box, type **N048*43'14.8065**, and press Enter.
5. In the East/West Lon. box, type **E002*21'37.8710**, and press Enter.
6. In the Altitude box, type **295**, and press Enter.
7. In the Heading box, type **78**, and press Enter.
8. Click OK to apply these settings.
9. Open the Sim menu and be sure the Auto Coordination option is checked. If it isn't, click on it or press A.
10. Open the Sim menu again, and select Realism and Reliability.
11. Set the Prop Advance to Fixed Pitch, and click OK.
12. Open the World menu and select Weather. Then select Edit.
13. In the Type box, select Clouds.
14. In the Base box, type **10000**, and press Enter.
15. In the Tops box, type **11000**, and press Enter.
16. In the Coverage box, select Scatter 3/8.
17. Click OK to return to the Weather dialog box. Now click OK again to apply these settings.
18. Open the World menu again, and select Set Time and Season.
19. Set the Season to Spring.
20. Set the Time of Day to Day.
21. Click OK to apply these settings.
22. Open the Options menu, and select Save Situation. Type **CH2FLT1** in the Situation Title box, press Enter, and click OK.

Before we depart, save these flight settings as a situation and name it CH2FLT1.STN. (See "Saving and Loading Situations," in the Introduction, for a refresher on how to do this.)

Adventures in Flight Simulator

Springtime in Paris

Well, this should be quite a change from Meigs Field in Chicago. The scenery here in Paris is very pleasant. We'll have a few moments to take in the scenery, but for the most part, you'll need to pay attention to what is going on in the cockpit. Before we get started, let's go over a few of the instruments that we'll be using during this lesson.

Basic Flight Instruments
Airspeed Indicator

One of the primary instruments for this flight is the airspeed indicator, which measures the aircraft's speed through the air in knots. In a real aircraft this information is gathered directly, through a small tube—called the Pitot tube—on the frontal surface of one of the wings. The airflow creates pressure in a static air line, which pushes on a diaphragm inside the instrument, which in turn causes the needle to move. Note that the airspeed indicator does not measure the speed of the aircraft relative to the ground. Thus, if an aircraft is flying at an airspeed of 140 knots with a 10-knot headwind, the ground speed is actually 130 knots.

The airspeed indicator has four colored areas that show the various operating ranges of the aircraft. The white arc shows the full flap operating range. The lower limit of the white area is the stall speed of the aircraft when landing. The upper limit of the white arc is the maximum airspeed permissible with full flaps extended. The green arc shows the normal operating range of the aircraft. The upper limit of the green area is the maximum cruising speed that can be attained without causing stress to the structure of the aircraft. The yellow area indicates caution, meaning that the aircraft can be operated only in smooth air at the indicated airspeed. Sudden turbulence at this airspeed can cause structural damage to the aircraft. The red line indicates the maximum airspeed for any type of operation with the aircraft. Exceeding the speed indicated by the red line turns the airplane into a collection of nonflying objects.

Airspeed indicator.

Altimeter

The altimeter measures the height in feet above sea level by reflecting changes in atmospheric pressure. Because changes in barometric pressure caused by changes in the weather can affect the accuracy of the altimeter, pilots must often adjust it to reflect the barometric pressure in their area. The small window on the right side of the instrument contains a barometric scale, and a knob for adjusting the instrument is located on the front. So when the tower or ATIS informs you of the barometric pressure, remember to adjust the altimeter for an accurate setting. Remember, also, that the height above sea level is not necessarily the same as ground level, so you will need to know the exact altitude of the airport you're using, and adjust the altimeter for it. The large hand of the altimeter indicates hundreds of feet, the small hand indicates thousands of feet, and the "T" indicates tens of thousands.

Directional Gyro

Also called a heading indicator, the directional gyro shows the heading of the aircraft. A digital readout above the aircraft image on the instrument shows the "To" heading. The readout below the aircraft image shows the "From" heading. The outside markings on the instrument indicate the full directional range around the aircraft.

Turn Coordinator

The turn coordinator measures the turn rate and helps the pilot to coordinate the turns of the aircraft. The airplane symbol on the instrument reflects the attitude of the aircraft by means of an electric gyro. When the wingtip of the airplane symbol

lines up with the L or the R, the aircraft is in a standard-rate, or two-minute, turn. The little ball indicates the coordination of the turn. We'll discuss this concept later, when we start using the controls without Auto Coordination to fly the aircraft.

From left to right: altimeter, directional gyro, turn coordinator, and vertical speed indicator.

Vertical Speed Indicator

This instrument is also called the rate-of-climb indicator. It shows the rate at which the aircraft is climbing or descending in hundreds of feet per minute. When the indicator is pointed at the number 5 above the 0, the rate of climb is 500 feet per minute. If it is pointed at the number 5 below the 0, the rate of descent is 500 feet per minute. Because this instrument relies on an internal diaphragm that moves in response to the air pressure outside the aircraft, the instrument does not immediately reflect the actual movement of the aircraft. In a real aircraft, the delay can be as great as 9 seconds. I have noticed some delay in the response of the simulated instrument, but it is not as pronounced as in real flight.

Chapter 2: Pilot Training and Private Lessons

Time For Takeoff

The tower has given us clearance for departure, so give it full throttle, and then gently pull back on the yoke at 80 knots. Raise the gear when the altimeter reads 100 feet.

Take Off

1. Release the parking brakes by pressing . (period).
2. Hold down keypad 9 to apply full throttle.
3. Keep the plane centered down the runway and rotate, or pull back on the yoke for liftoff, at 80 knots.
4. At 100 feet, press G to raise the gear.

Try to keep the airspeed at around 80 knots, but don't worry about going over. The important thing is not to let it drop below 80 knots. This will give us a good rate of climb as we head for a target altitude of 1500 feet. Go ahead and look out the back of the plane toward the runway for a good view of the airport. Now look forward again, and check the airspeed indicator to be sure that the airspeed is still above 80 knots. Look at the vertical speed indicator, and notice the rate of climb.

You will also notice that in order to maintain this climb attitude, you have to keep constant back pressure on the yoke. To help alleviate the need for this constant pressure, use the elevator trim to adjust the pitch of the aircraft. You might also want to look out the side windows and check the angle of the wings in relation to the horizon. Return to the forward view and keep an eye on the airspeed and on the altimeter. During our flight we must remember to always keep an eye out for other aircraft traffic, especially before making a turn.

Climb Out

1. Apply forward or back pressure on the yoke to maintain an airspeed of at least 80 knots.
2. Press keypad 1 and keypad 7 to use the elevator trim to adjust the pressure needed on the yoke.
3. Press Shift-keypad 2 to look back at the runway.
4. Press Shift-keypad 8 to return to the forward view, and watch the airspeed and vertical speed instruments.
5. Press Shift-keypad 4 and Shift-keypad 6 to look out the side windows, and notice the wing angle.
6. Press Shift-keypad 8 to return to the forward view.

Straight and Level

Let's get ready to level off the aircraft. The procedure for doing this is to drop the nose of the aircraft, wait for the airspeed to increase, gradually decrease the throttle for straight and level cruising, and then use the elevator trim to neutralize the pressure needed on the yoke. We'll level off when the plane reaches 1500 feet. Remember to look out one of the side windows to be sure that the bottom line of the wing is parallel with the line of the horizon, indicating that we are in level flight. As the airspeed increases, the aircraft will want to start climbing again. When you feel (or see) this happening, use pressure on the yoke to keep the nose down; then trim it out again with the elevator trim. This will take some getting used to, and you might have to do it a few times until the airspeed stabilizes at around 135 knots. But don't worry about not being able to level off at exactly the right altitude. Once we have the plane flying straight and level, we can make the adjustments necessary for flying at the correct altitude.

Level Off at 1500 Feet

1. At 1500 feet, apply forward pressure on the yoke to drop the nose for level flight.
2. Wait for the airspeed to build up, and then press keypad 3 several times to reduce the throttle to 2300 RPM.
3. Press Shift-keypad 4 to look to the left to check the wing attitude for level flight.
4. Press Shift-keypad 8 to return to the forward view.
5. Use keypad 1 or keypad 7 (the elevator trim control) to trim the aircraft for level flight at 1500 feet.

As the aircraft approaches an altitude of 1500 feet, we should hear the beeping of the middle marker beacon. We can tell that it is the middle marker from the flashing *M* light on the radio panel. I will talk about using the beacons when we start the flying lessons that cover using the navigational instruments. However, we will encounter the outer marker beacon shortly.

Avoid "chasing the needle" on the vertical speed indicator when you're trying to fly straight and level. Remember that the instrument is delayed in responding to the actions of the aircraft. Some pilots get in the bad habit of using the vertical speed indicator to determine whether they are flying at a constant altitude and then end up correcting for the climb or descent when they see the needle move. This can cause

Chapter 2: Pilot Training and Private Lessons

Aeronautically Speaking:
When two aircraft in an uncontrolled airspace come within a mile of each other, it is called a *near miss*. (I always felt that the term *near hit* would be more appropriate.) Uncontrolled airspace is any area in which the aircraft are not under direct radio and radar contact with air traffic control (ATC).

excessive pitching of the aircraft. Once you have stabilized at the target altitude, it is okay to use this instrument as a backup to the altimeter to confirm that the plane is in level flight. But your primary instruments for straight and level flight should be the left and right windows (for looking at the wings) and the front window (for noticing the pitch of the nose in relation to the horizon). You can use the altimeter and the vertical speed indicator as backups.

Standard-Rate Two-Minute Turns

Because we have not made any turns yet, our heading should still be at 77°. (Even though we entered a heading of 78° at the beginning of this flight, Flight Simulator showed a compass heading of 77° due to its mathematical rounding.) About this time we should be hearing the outer marker beacon tone. At the same time, the *O* light on the radio panel will be on. We will use this occurrence as a starting point for a standard-rate turn to the left. But first, let's visually check the airspace that we will be turning into to be sure that there are no other aircraft in that airspace or headed toward that airspace.

After we've checked the area by looking left front, left, and left rear, we can begin our standard-rate turn, which is also called a *two-minute turn*. While in a standard bank, an aircraft takes approximately two minutes to complete a full 360° turn.

Keep an eye on the turn coordinator, and bank the plane to the left until the wingtip of the airplane on the instrument is pointed at the index mark above the letter *L*. Remember to apply a little back pressure to the yoke to keep the plane level at 1500 feet. During the turn, the order of your scan should be as follows: left window, front window, turn coordinator, airspeed indicator, altimeter, and then back to the left window. The aircraft will have a tendency to gradually come out of the turn, so you will need to adjust the pressure on the yoke to maintain the required bank. As we near a heading of 280°, begin to come out of the turn for a final heading of due west, or 270°, and return to straight and level flight. Again, make whatever corrections are necessary to fly the heading of 270° at an altitude of 1500 feet.

Make a Two-Minute Turn to the Left

1. Maintain straight and level flight until you hear the tone of the outer marker beacon.
2. Visually check the airspace left of the plane by pressing Shift-keypad 1, Shift-keypad 4, and Shift-keypad 7. Then press Shift-keypad 8 to return to the forward view.
3. Apply left pressure to the yoke, and bank until the turn coordinator shows a standard bank angle for a two-minute turn.
4. Apply back pressure to the yoke as necessary to maintain your current altitude.
5. Press Shift-keypad 4 to see the left view; then press Shift-keypad 8 to return to the forward view, and scan the turn coordinator, the airspeed indicator, and the altimeter.
6. At a heading of 280°, start returning to straight and level flight at a final heading of 270°.

Power Climb

Good! You have just done a standard-rate turn. Now we should be heading due west, with a heading of 270°, and we should be able to see a river up ahead. About this time we can expect to hear the outer marker beacon tone again. This time we will ignore it and continue with our lesson.

Our airspeed should be between 120 and 130 knots. Now let's see what effect the throttle has on the pitch of the aircraft. We'll do another climb, but this time we'll use the throttle. Our target altitude for this climb will be 3000 feet. Apply full

Wing Tip:
When you're preparing to come out of a turn, remember to start leveling the wings 10° before the target heading. When you are turning left, this would be the target heading plus 10°, or 280°, for a target heading of 270°. When you are turning right, this would be the target heading minus 10°, or 80°, for a target heading of 90°.

Wing Tip:
If you are having a difficult time trying to fly straight and level, remember that it does get easier with practice. Be gentle with the controls, and move smoothly, instead of abruptly, into the attitude that you are trying to hold. It is easy to overcompensate for sudden movements, so less movement is better than too much.

You can configure Flight Simulator for different sensitivities with the controls. To do this, open the Options menu and select Preferences. Select the Keyboard, Mouse, or Joystick button to display the necessary controls for adjusting sensitivity. Having a less sensitive control will help you learn to handle the aircraft, but the higher the sensitivity setting, the more realistic the controls.

throttle, and notice how the airplane begins climbing on its own, without our having to touch the yoke.

Now that we see how the throttle affects the lift or lack of lift, let's continue to climb to the target altitude. Gently pull back on the yoke to increase the rate of climb. Notice that the vertical speed indicator shows the increased rate of climb. Keep a close eye on the airspeed to make sure that it does not drop lower than 80, or you'll end up getting a quick lesson in stalls. After you reach 3000 feet, level off the aircraft. Double-check to make sure that the heading is still at 270°. As you drop the nose of the aircraft, the airspeed will build up again, and the plane will start to climb. Adjust the pitch with the elevator trim as needed. Press P to pause if you want to rest a little.

Climb to 3000 Feet

1. Hold down keypad 9 to apply full throttle. Notice how the aircraft wants to climb.
2. Apply back pressure on the yoke until the airspeed is around 80 knots, and climb to 3000 feet.
3. Release pressure from the yoke to level off at 3000 feet, keeping a heading of 270°.
4. Use keypad 1 and keypad 7 to trim the plane for straight and level flight.

Sight-Seeing

Now that we are back to straight and level flight, let's take some time to look around a little. Our heading should still be 270°, due west, and the airspeed should be back up to around 140 knots. We've finished our climb, so we can reduce the throttle to 2300 RPM. You will need to trim the airplane to compensate for the reduction in the power setting. You should be able to see Paris and the Eiffel Tower coming up on the right. If you don't see them yet, keep flying straight and level until they appear directly off the right side of the airplane. Switch to Spot view, which will give you a better look at the whole area. Then change the view to the left side of the aircraft.

When you're looking at scenery, remember to check the instruments and the attitude of the aircraft regularly. Check the positions of the nose and wings in relation to the horizon. You can also look out the back and off to the left, if you want, and check out the surrounding scenery. Notice the roads and the rivers around the area.

Fly Straight and Level for Some Sight-Seeing

1. Maintain straight and level flight at 3000 feet.
2. Press keypad 3 several times to reduce the throttle to 2300 RPM.
3. Use keypad 1 or keypad 7 (the elevator trim control) to trim the aircraft.
4. Press Shift-keypad 6 to change the view. Look for Paris and the Eiffel Tower.
5. Press S twice to switch to Spot view.
6. Press Shift-keypad 4 to change to a view of the left side of the aircraft.
7. Look around at the different views. (Press Shift and any key on the numeric keypad.)
8. After you pass Paris, press S to return to Cockpit view.
9. Press Shift-keypad 8 to switch to the forward view from the cockpit.

Continue with straight and level flight until the city appears under the right wing. It will be off in the distance, and it might be hard to see the buildings, but you can distinguish a gray circle surrounded by a green one.

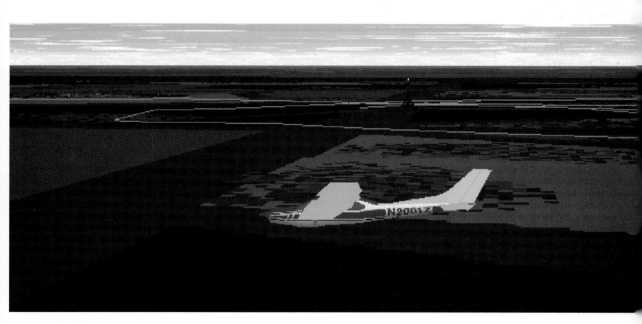

Chapter 2: Pilot Training and Private Lessons

After the city appears under the right wing, make a standard-rate turn to the right. Remember to have the right wingtip of the aircraft image on the turn coordinator pointed to the index mark above the letter *R*. Do this while maintaining a constant altitude of 3000 feet. End the turn at a heading of 000°, or due north. Remember to start coming out of the turn at 350°, 10° before the desired heading.

After returning to straight and level flight, you should quickly check the engine instruments to be sure that there is plenty of fuel and that the oil pressure and temperature readings are still in the green area.

Make a Standard-Rate Turn to a Heading of 000°

1. Apply right pressure on the yoke to bank until the turn coordinator shows a standard bank angle.
2. Use a little back pressure on the yoke to maintain an altitude of 3000 feet.
3. Begin leveling out the aircraft at a heading of 350° to end the turn at a heading of 000°.
4. Return to straight and level flight.
5. Check the engine instruments for fuel levels, oil temperature, and oil pressure. (You will need to press Tab if you are not in SVGA mode.)
6. Flip back to the communications instruments if necessary. (You will need to press Tab to do this if you are not in SVGA mode.)

Are you starting to feel comfortable with the controls and the aircraft? If not, you can always start this flight over and try it again. You can also save the situation whenever you want so that you can come back and continue from that point in the flight. But if you're ready, let's continue.

You should now be headed toward the city buildings just outside of the Boulevard Peripherique, the circular road that surrounds the metropolitan area of Paris. Just off the right of the aircraft should be the river Seine, which runs through the middle of Paris and right next to the Eiffel Tower. Feel free to look around a little as we maintain straight and level flight. Remember to check the altimeter, the turn coordinator, and the directional gyro, and to check the attitude of the aircraft visually to maintain an altitude of 3000 feet and a heading of 000°. I realize that I am repeating these basic instructions, but it is very important to acquire the habit of checking the condition and attitude of the aircraft.

After returning to straight and level flight, look out the right window to be sure that we are coming up on the Eiffel Tower. Due to the limiting view out the right side of the cockpit it may be necessary to use Spot view from the left of the aircraft.

View the Eiffel Tower

1. Press S twice to switch to Spot view.
2. Press Shift-keypad 4 to position the view from the left of the aircraft.
3. Press P to pause the situation just before the Eiffel Tower appears to reach the right wing of the aircraft.

Turns Around a Point

We are getting ready to do "a turn around a point." Using a ground reference, you'll fly in a coordinated circle around an object on the ground. In this case, the object will be the Eiffel Tower. The trick is to keep the wing of the aircraft pointed at or just above the object on the ground while maintaining a constant altitude. Now would be a good time to save the situation. Then if you encounter some difficulties with this maneuver, you can keep trying it until you become more comfortable with it.

Unpause Flight Simulator when you are ready to begin doing the turn around the point, and return to Cockpit view. Look out the right window and wait for the Eiffel Tower to be lined up directly under the right wing. Then bank the aircraft to the right until the bottom of the wing is just above the Eiffel Tower. If the Eiffel Tower appears to move behind the wing, you need to increase the angle of the bank. If the wing appears to fade back behind the Eiffel Tower, you need to decrease the angle of the bank. The closer you are to the object that is the focal point, the steeper the bank must be for a full coordinated turn around the point.

Remember to apply a little back pressure to the yoke to maintain a constant altitude. Don't worry if you lose or gain a little altitude, as long as the change is not too drastic. You might want to check the altimeter to be sure that you're at 3000 feet.

Buzzing the Eiffel Tower.

This will be a one-and-one-quarter turn around the Eiffel Tower. We started the turn at a heading of 000°, so we will fly a full circle back to 000° again; but instead of coming out of the turn at that point, we will continue turning until we reach a heading of 90°, due east. Because we are not close to the Eiffel Tower, our bank will be shallow, and making the complete turn will seem to take forever. But it will give you some good practice in maintaining a constant turn around a fixed object on the ground.

Make a Turn Around the Eiffel Tower

1. Press P to unpause Flight Simulator, and then press S to return to Cockpit view.
2. Press Shift-keypad 6 to look out the right window. (You might have to bank a little to see the Eiffel Tower because of the limited view out the right window.)
3. When the wing is over the Eiffel Tower, apply right pressure to the yoke, banking the plane until the wing is pointed just over the top of the tower.
4. Maintain this bank through a full circle around the Eiffel Tower, decreasing the angle of the bank if the tower appears to move in front of the wing, and increasing the angle of the bank if the tower appears to move behind the wing.
5. Use back pressure on the yoke to maintain an altitude of 3000 feet.
6. After reaching a heading of 000°, continue toward a heading of 90°, and start coming out of the turn at 80°.
7. Press Shift-keypad 8 to return to the forward view.
8. Fly straight and level at a heading of 90°, with an altitude of 3000 feet.

Coming out of this turn should be quite easy because the bank was so shallow. Make whatever corrections you need, however, to get the aircraft on the correct heading and altitude.

While you continue east flying straight and level, look at the various sights and see whether you spot the Eiffel Tower out the back of the aircraft. Use Spot view for a better glimpse of the scenery. If you see something special, go ahead and take a "snapshot" of it.

Switch to Spot View and Take a Flight Photograph

1. Press S twice to switch to Spot view.
2. Press Shift and any of the cursor keys on the numeric keypad to view your surroundings. Find a view that you like.
3. Open the Views menu, and select Flight Photograph.
4. Enter a filename of up to eight characters in the Filename box, and press Enter.
5. Click OK to save the screen as a PCX bitmapped picture.

Wing Tip:
The Flight Photograph function creates a PCX bitmap file. You can view a PCX file with Microsoft Paintbrush in Microsoft Windows, or you can view it with one of many commercial or shareware graphics packages. Flight Simulator saves these files in its root directory.

Tower Buzz

Now that you have studied standard-rate turns, climbs, turns around a point, and turns to a heading, how about a little fun? I know you're just dying for a chance to "buzz" the Eiffel Tower, so let's do a standard-rate left turn to a heading of 000°, due north. After the standard-rate turn, reduce the throttle to full off and drop the nose for a descent rate of around 100 to 120 knots. If the aircraft is not pointed directly at the Eiffel Tower, compensate as needed by turning until the nose is pointed slightly to the left or right of the tower.

We want to plan our descent rate so that the aircraft is around 500 feet above ground level when we reach the Eiffel Tower. Using your best judgment, try to adjust the rate of descent to achieve this goal. If you reach 500 feet before you reach the tower, apply 50-percent power and trim the aircraft for straight and level flight.

Prepare to Buzz the Eiffel Tower

1. Apply left pressure to the yoke until the turn coordinator shows a standard bank angle.
2. Start coming out of the turn at 10° for a target heading of 000°.
3. Look for the Eiffel Tower as you fly straight and level. (Use Shift and the numeric keypad keys to help you locate the tower if necessary.)
4. Turn as needed to aim the aircraft slightly left or right of the Eiffel Tower.
5. Hold down keypad 3 until you cut power completely.
6. Adjust pressure on the yoke to descend at a speed of 100 to 120 knots. Try to be at 500 feet above ground level (AGL) just before we arrive at the Eiffel Tower.
7. Press keypad 9 several times to increase power to 50 percent if necessary for straight and level flight to the Eiffel Tower.

Moving along at over 100 knots, you will buzz past the Eiffel Tower quite rapidly. Just before you reach the tower, switch to Spot view. If the tower will be on the right of the aircraft, set Spot view for the left side, or vice versa if the tower is on the left side of the aircraft. As you fly by the tower, dip the wing that is closest to the tower and give the tourists a bit of a thrill. After you pass the tower, pause the simulation, and I'll show you how we can relive the event.

Buzz the Eiffel Tower with Spot View

1. Press S twice to switch to Spot view just before we arrive at the Eiffel Tower.
2. Use Shift and the numeric keypad keys to adjust the Spot view as necessary to view the aircraft and the Eiffel Tower.
3. Press P to pause the flight after we pass the Eiffel Tower.

Wing Tip:
Instant Replay can come in handy for reviewing those spectacular crashes.

How many pilots get a chance to buzz the Eiffel Tower in a real aircraft? Not many! Would you like to see a replay of the buzzing of the tower? You can do that using a Flight Simulator command called Instant Replay. Go ahead and select Instant Replay; use the default settings for viewing the segment. After you watch the segment at the regular speed, you can go back and watch it again in slow motion. This time, change the settings for a slow viewing rate. Try 50 or 25.

View the Instant Replay

1. Open the Options menu, and select Instant Replay.
2. If the Repeat Replay box is not checked, select this option by clicking on the box.
3. Click OK.
4. Watch the replay, and then click OK.
5. When the Instant Replay dialog box appears after the replay, type **50** or **25** in the Replay Speed box, and press Enter.
6. Click OK.
7. Watch the replay, and click OK again.

This marks the end of the first lesson. In this flight, we covered the following topics:

The takeoff
Climbs to a target altitude
Straight and level flight at a target altitude
Standard-rate, two-minute turns to a target heading
Turns around a point
Flight photographs
Buzzing the Eiffel Tower
The Instant Replay feature

Chapter 2: Pilot Training and Private Lessons

Preparing for Landings

In the first 10 to 20 hours of professional flight training, about 50 percent of the time is spent in preparing the pilot for landings. Compared to landing, flying an aircraft is actually quite easy. Any pilot will tell you that the most work-intensive function of flying is getting the plane back on the ground, preferably on a runway.

In preparation for learning how to land, we can perform some related maneuvers that will help you become comfortable with how an aircraft behaves during a landing and teach you how to control it.

Flight Plan: This flight will be in the Cessna R182 over the Rocky Mountains in the western United States. Because you should already be comfortable with taking off, we will begin in the air. During this flight, we will discuss and practice power-off stalls, power-on stalls (at 30-percent power), and slow flight. We'll concentrate on how the aircraft "feels" in the slow-flight situations that occur during landing approaches. We won't use ground references but will rely on the basic instruments.

Before you enter the flight settings, be sure to pause Flight Simulator. As you will see shortly, this is very important. After you enter the settings, save the situation.

Flight Settings

Location	Over the Rockies
Aircraft	Cessna Skylane RG R182
North/South Latitude	N042°16'01.4871
East/West Longitude	W106°29'15.5242
Altitude	7000 feet
Heading	180°
Auto Coordination	On
Season	Winter
Time of Day	6:30 a.m.
Weather—Clouds	
Base	10,000 feet
Tops	15,000 feet
Coverage	6/8, Broken
Realism and Reliability	
Prop Advance	Fixed Pitch

Enter the Flight Settings

1. Press P to pause Flight Simulator.
2. Open the Options menu, and select Aircraft.
3. Select the Cessna Skylane RG R182, and click OK.
4. Open the World menu, and select Set Exact Location.
5. In the North/South Lat. box, type **N042*16'01.4871**, and press Enter.
6. In the East/West Lon. box, type **W106*29'15.5242**, and press Enter.
7. In the Altitude box, type **7000**, and press Enter.
8. In the Heading box, type **180**, and press Enter.
9. Click OK to apply these settings.
10. Open the World menu and be sure Slew is not checked.
11. Open the Sim menu, and be sure there is a check mark in front of Auto Coordination. If it is not selected, click on the option or press A.
12. Open the Sim menu again, and select Realism and Reliability.
13. Set Prop Advance to Fixed Pitch, and click OK.
14. Open the World menu, and select Weather. Then select Edit.
15. In the Type box, select Clouds.
16. In the Base box, type **10000**, and press Enter.
17. In the Tops box, type **15000**, and press Enter.
18. In the Coverage box, select Broken 6/8.
19. Click OK twice to apply these settings.
20. Open the World menu again, and select Set Time and Season.
21. Set Season to Winter.
22. Select Set Exact Time, type **6** in the Hours box, and press Enter. Type **30** in the Minutes box, and press Enter.
23. Click OK to apply these settings.
24. Press L if necessary (or click on the lights indicator) to turn on the instrument lights.
25. Open the Options menu, and select Save Situation. Type **CH2FLT2** in the Situation Title box, and press Enter.
26. Click OK to save this situation.

Chapter 2: Pilot Training and Private Lessons

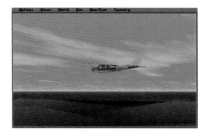  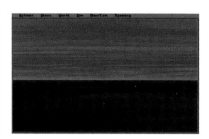

Over the Rockies

Ever been flying over mountains? They can be very beautiful, especially when covered with snow. During this lesson, we'll be flying over the Rocky Mountains. So let's get going.

Before you unpause Flight Simulator, you should know that you cannot set the throttle via a preference or a menu setting. Therefore, although the aircraft is starting at an altitude of 7000 feet, the throttle is off and the gear is down. You can go ahead and increase the throttle while the simulator is paused, but you will have to wait until it is unpaused before you raise the gear. Unpause the simulator and raise the gear. The nose of the aircraft will drop immediately because we started this situation in mid-air with no speed. Gently pull back on the yoke and trim the aircraft for straight and level flight at 6000 feet.

Start the Flight

1. Hold down keypad 9 to apply full throttle.
2. Press P to unpause Flight Simulator.
3. Press G to raise the gear.
4. Pull back on the yoke to recover from the initial dive once you have gained an airspeed of 80 knots.
5. Maintain straight and level flight at 6000 feet, with a heading of 180°.

Owing to the early hour, it is still a little dark outside. The stars are still visible and the clouds are a delicate shade of pink. The instrument panel is red from the glow of the night-flight lights. As we progress through our flight, the sky will get lighter, the stars will begin to fade, and the red lights on the instrument panel will go off.

As you get comfortable with the airplane again, look around at the scenery a little. Enjoy the various views from inside the cockpit, and then switch to Spot view. This reminds me of the Denver, Colorado, area: mountains to the west, city to the east, and then flat for the next thousand miles east. Bring the throttle back to about 80 percent. Remember to refer to the vertical speed indicator, altimeter, directional gyro, and airspeed indicator to ensure that the aircraft remains straight and level.

Maintain Straight and Level Flight at 80-Percent Throttle

1. Press Shift and any numeric keypad key to look at the various views from the cockpit.
2. Press S twice to switch to Spot view.
3. Press Shift and any numeric keypad key to look at the various external views.
4. Press S to switch back to the Cockpit view.
5. Press keypad 3 several times to reduce the throttle to around 80-percent power.
6. Check the instruments and adjust the aircraft to straight and level flight, applying pressure to the yoke and using the elevator trim (keypad 1 and keypad 7).

Slow Flight

The first procedure we execute is slow flight. During slow flight, we will continue to fly the aircraft at reduced throttle. Because less air passes over the wing than it does when we are at full throttle, the handling of the aircraft will be sluggish; turning and banking maneuvers will be slower. The first part of our slow flight will be at a throttle setting of 50 percent. We'll fly straight and level and do a few turns at this setting. Then we'll reduce the throttle to 30 percent and continue to fly straight and level.

Wing Tip:
It's easy to get involved with reading a lesson while the aircraft continues to fly. When you return your attention to the aircraft, you might find that it has gone beyond a point that is referenced in the lesson. Feel free to pause the simulation during the lessons and flight adventures anytime you need to. After flying through a lesson or adventure, you can always come back to the flight and go through it again without having to reread the entire lesson or description. Just follow the action steps.

Aeronautically Speaking:
If there are a lot of bright lights in the cockpit during night flights, it is hard for the pilots to adjust their vision so that they can see into the night sky to check for other air traffic. The red lights on the instrument panel allow the eyes of the pilots to stay adjusted for night vision. Sometimes even the red lights can be too bright; an adjustment knob allows the pilots to change the lights' intensity. During the first hour of a night flight, pilots will gradually reduce the brightness of the red lights as their eyes adjust to the darkness.

Let's get started by reducing the throttle to 50 percent. As you reduce the throttle, you'll need to apply back pressure on the yoke to compensate for the reduction in the lift, and you'll need to point the nose a little higher above the horizon. As the aircraft begins to slow down, the nose tends to drop. Apply a little more back pressure to the yoke as needed. You can use the elevator trim to stabilize the back pressure on the yoke. It's okay if you lose a little altitude while you try to adjust to this new attitude. But the tops of the mountains are at around 4000 feet, and it would be a good idea to stay at least 1000 feet above the mountains, so don't lose too much altitude! You'll have to continue to make adjustments until the aircraft finally stabilizes at a constant altitude of 6000 feet. The airspeed should be around 105 knots.

Practice Slow Flight at 50-Percent Throttle

1. Press keypad 3 several times until the throttle is at 50-percent power (in the middle of the throttle control).
2. Gently apply back pressure on the yoke to maintain straight and level flight.
3. Use the elevator trim (keypad 1 and keypad 7) to adjust the aircraft to a new attitude.
4. Stabilize the aircraft in straight and level flight at an airspeed of around 105 knots.

Now that you have become comfortable with the 50-percent power setting (or at least I hope you have!), we'll climb a bit. Add a little back pressure on the yoke, and try to maintain an airspeed of around 80 knots. Be very careful not to let the airspeed drop below 80 knots. Even though we are still 20 knots above stall speed, it's a good idea to leave a little margin to ensure the maneuverability of the aircraft. You will notice that at this throttle setting, our climb rate is a lot less than it is at full throttle. Continue the climb to an altitude of 6500 feet. This should give us enough distance above the mountains for our next set of maneuvers. When we reach 6500 feet, level off, but keep the throttle at 50 percent. One difference between this climb and the climbs we did earlier is that we do not need to drop the throttle after leveling off. The reduced airspeed makes the transition between the climb and straight and level flight a little easier because the slower movements of the aircraft are easier to compensate for.

Climb to 6500 Feet and Return to Straight and Level Flight

1. Apply back pressure to the yoke to put the aircraft in a climb attitude.
2. Keep the airspeed above 80 knots during the climb, and level off at 6500 feet.
3. Return to straight and level flight.

Now that we are back to straight and level flight, let's check the position of the wings in relation to the horizon. In the introductory flight, we used the wings to visually check for straight and level flight, but that technique doesn't apply during slow flight. Press Shift-keypad 4 to look out the window to the left and notice how the wing of the aircraft points toward the sky instead of lying parallel with the horizon. The aircraft looks like it is climbing, even though it is flying straight and level, because of its decreased airspeed. Because less air is passing over the top of the wing, the angle of the wing must change in order to maintain the lift needed to stay at a constant altitude.

Slower Flight

Let's go on now, and do a slow flight at 30-percent power. Reduce the throttle to 30-percent power (about a third of the way from the bottom). The RPM should be around 1800. As the nose of the aircraft starts to dip, compensate by applying back pressure to the yoke. You can also use the elevator trim to adjust the aircraft to the new attitude.

Your new target airspeed should be about 85 knots. At that speed, it might take awhile to get the aircraft to stabilize in straight and level flight. Instead of using visual clues, rely on the instruments to achieve the attitude. Again, try not to let the airspeed drop below 80 knots. As we get closer to 85 knots, the nose of the airplane will start to drop a little. Let it drop. You will find that as the airspeed increases by a couple of knots, the nose will come back up again. This rising and falling movement is called *porpoising*, after the motion of a porpoise as it repeatedly leaps above the water and then dives down again. These changes in pitch gradually lessen as the aircraft stabilizes in straight and level flight.

Aeronautically Speaking: The angle of the wing in relation to the airflow is called the **angle of attack**. As the airflow decreases, the angle of attack must be increased to maintain the amount of lift. If the angle of attack becomes too steep, the airflow over the wing will be disrupted and will no longer provide lift, resulting in a stall. The angle at which a stall would occur is called the **critical angle of attack**.

Chapter 2: Pilot Training and Private Lessons

Practice Slow Flight at 30-Percent Throttle

1. Press keypad 3 several times until the throttle is at 30-percent power (a third of the distance from the bottom).
2. Apply back pressure to the yoke to maintain level flight.
3. Use keypad 1 and keypad 7 as needed to apply elevator trim.
4. At about 85 knots, let the aircraft porpoise a little as it stabilizes.

Flying at 30-percent throttle conserves fuel, but it is also very slow. Compared to cruising at a speed of 156 knots (at 7500 feet), it would take almost twice as long to reach a destination at 30-percent power. In the real world, it is also not a very safe way to fly because the aircraft is so sluggish at that speed that a sudden gust of turbulence could cause a wing to stall, resulting in a spin. Fortunately, we don't have to worry about that in Flight Simulator.

Let's take another look at the wing. Notice the change in the angle of attack in relation to the horizon. Again, this looks as if the aircraft were climbing instead of flying straight and level. Let's take a look from the Spot view from the left side of the aircraft. Our Spot view has us looking down on the aircraft at an angle. We need to change to a view that is level with the aircraft so that we will be able to see the angle of attack in the aircraft during very slow flight.

View the Cockpit and Spot Views

1. Press Shift-keypad 4 to look out the left window, and notice the angle of the wing in relation to the horizon.
2. Press S twice to switch to Spot view.
3. Press Shift-keypad 4 to set the Spot view to the left side of the aircraft.
4. Open the Views menu, and select Set Spot Plane.
5. In the Altitude box, type **0**, and press Enter.
6. Click OK to apply this setting.

We can see from this view that the aircraft is in a nose-up attitude in relation to the horizon. Because the scenery here is spectacular, it would be a good time to take a flight photograph.

While we are looking at the aircraft in Spot view, we can also adjust the zoom. By pressing the plus and minus keys we can get closer to the aircraft or move out for a distant view. I prefer to view the plane one zoom level out from the default setting of Flight Simulator. This is close enough to see details of the aircraft but far enough away to have a good view of the scenery as well.

Take a Flight Photograph and Use Zoom

1. Press Shift and any numeric keypad key to select the view that you want for your photograph.
2. Open the Views menu, and select Flight Photograph.
3. Type a filename of up to eight characters in the Filename box, and press Enter.
4. Click OK.
5. Adjust the view to your liking by using the – (minus) key at the top of the keyboard to zoom away from the aircraft or the + (plus) key at the top of the keyboard to zoom in closer.
6. Press S to return to Cockpit view.

Now that we have returned to Cockpit view, we can prepare for the next set of maneuvers for this lesson. It's okay if you lost a little altitude while you were in the external views. Unless you are looking at the instrument panel, it is a little difficult to determine whether the plane is climbing or descending.

Slow Turns

While we are flying at 30-percent power, let's make a turn to the left. We can't make a standard-rate turn because the bank angle for a standard-rate turn would be too steep for our reduced speed. For this turn we will be using the attitude indicator, the instrument just above the directional gyro. We will use the white index markings on the outer rim of the instrument for measuring our angle of bank during this turn.

We'll turn to a heading of 90° due east. We want to bank the aircraft to the left until the triangular pointer at the top of the attitude indicator (also called the artificial horizon) is pointing at the first index dot to the left of the center index mark. We'll want to maintain this bank angle throughout the turn. Start coming out of the turn at 100° for a target heading of 90°.

Make a Shallow Turn to the Left

1. Apply gentle left pressure to the yoke, banking until the attitude indicator triangle is pointing to the first dot on the left of the center mark.
2. Use a slight amount of back pressure on the yoke to maintain altitude.
3. Turn to a heading of 90°.

Attitude indicator.

The reason for such a shallow bank is that there is less lift during a turn than in level flight. Because lift is strictly a vertical force, any deviation of the wings from horizontal position decreases the amount of lift. The steeper the turn, the less lift the wings produce. High airspeeds counteract this decrease in lift, but at 30-percent power, we can't afford to bank steeply!

Let's make another shallow turn, but this time, we will make a turn to the right. Turn to a heading of 180°. Use the attitude indicator again, and bank the aircraft until the pointer of the instrument is pointing at the first index dot to the right of the center index mark. When you are finished with the turn, pause the simulator.

Make a Shallow Turn to the Right

1. Apply gentle right pressure to the yoke, banking until the attitude indicator triangle is pointing to the first dot on the right of the center mark.
2. Use a little back pressure on the yoke to maintain altitude.
3. Turn to a heading of 180°.
4. Press P to pause Flight Simulator.

The Dreaded Stall

In our next maneuver, we'll go into a stall while flying at 30-percent power. This is similar to a power-off stall. A stall is basically the same kind of maneuver as landing on a runway. When landing, the aircraft is stalled just a few inches above the runway. Practicing stalls will help you prepare for landings. In the stall maneuver we'll gradually apply back pressure to the yoke and bring the nose of the aircraft up until the wings can no longer provide lift. Just before we lose all lift, a warning buzzer will sound, letting us know that we are about to stall. We will maintain back pressure on the yoke until the nose of the plane actually drops toward the horizon. By ignoring the buzzer, we'll be able to "feel" what the stall is like. (If I had understood this maneuver before my first flying lesson, I would have been a lot more comfortable. See the Preface for my story.)

Okay, unpause the simulator and let's give it a try. Apply a little back pressure on the yoke. As the nose dips toward the horizon, apply more back pressure to the yoke. Try to keep the horizon right at the top of the instrument panel. We do not want to use the elevator trim for this procedure because we want to recover rapidly after the stall. Continue to pull back gradually on the yoke as the airspeed drops. The stall will occur at an airspeed of about 60 knots, but don't take my word for it.

You need to "feel" the stall. At 60, we will hear the stall warning buzzer. Continue to hold the yoke back until the plane dips toward the horizon. As soon as this happens, lower the nose and let the airspeed build back up to our straight and level flight speed for this power setting, which should be around 80 knots.

Chapter 2: Pilot Training and Private Lessons

Perform a 30-Percent Power Stall

1. Press P to unpause Flight Simulator.
2. Gently apply back pressure on the yoke so that the horizon appears to be at the top of the instrument panel.
3. As the nose starts to dip, apply more back pressure to the yoke and maintain it until the stall buzzer has sounded.
4. When the nose suddenly pitches down at around 60 knots, lower the nose, neutralize pressure on the yoke, and let the airspeed build back up to 80 knots.
5. Maintain straight and level flight.

A Stall in Landing Configuration

Let's do another stall. But this time we'll do a power-off stall in the same configuration that we would normally have during a landing. To get started, we need to turn on the carburetor heat and lower the flaps to 10°. Doing this increases the lift of the aircraft. Compensate for the added lift by adjusting the yoke to maintain straight and level flight. Even though we will attempt to maintain straight and level flight, the aircraft will start to descend at a gradual rate. This is okay because it simulates what will happen during an approach to a runway. As long as the airspeed registers in the white arc on the airspeed indicator, it is okay to lower the flaps and the gear. Lowering the flaps will decrease the airspeed. Lower the flaps to 30°. Again, you will need to compensate for the increase in the lift. After stabilizing the aircraft, lower the gear. The airspeed will again drop. Continue to adjust for straight and level flight. Lower the flaps to their full 40° and reduce the throttle to 0-percent power. Apply more back pressure on the yoke, and try to keep the plane in a nose-up attitude. You want to apply enough back pressure on the yoke to stop the gradual descent and bring the plane into straight and level flight.

You'll notice that as we pass through an airspeed of 60 knots, the aircraft is still flying. Thanks to the flaps, the lift has been increased. As we near 50 knots, the stall warning buzzer will begin to sound. Maintain the back pressure on the yoke until the nose of the aircraft drops to the horizon. As soon as this happens, release the back pressure on the yoke and apply full throttle. As the airspeed builds back up, raise the flaps fully, turn off the carburetor heat, and raise the gear. Then return to straight and level flight, with a throttle setting of 80 percent. After you stabilize the aircraft in straight and level flight, pause Flight Simulator.

Perform a Power-Off Stall in Landing Configuration

1. Press H to turn on the carburetor heat.
2. Press F6 to lower the flaps to 10°.
3. Apply pressure to the yoke as necessary to maintain straight and level flight.
4. Press F7 to lower the flaps to 30°. Compensate for the new attitude, and maintain straight and level flight.
5. Press G to lower the gear, and continue straight and level flight.
6. Press F8 to lower the flaps to 40°.
7. Hold down keypad 3 to reduce the throttle to 0-percent power. Continue with back pressure on the yoke for straight and level flight until the stall buzzer has sounded.
8. When the nose pitches down at about 50 knots, lower the nose, neutralize pressure on the yoke, and hold down keypad 9 to apply full throttle.
9. Press H to turn off the carburetor heat.
10. Press G to raise the gear.
11. Press keypad 3 several times to reduce the throttle to 80-percent power.
12. Return to straight and level flight.

Aeronautically Speaking: *The use of flaps changes the shape of the wings and produces more lift, thus allowing the aircraft to stay aloft at slower speeds.*

How was that for being busy? Just imagine doing this while keeping an eye on the location of the runway and then stalling the aircraft about six inches above the pavement. Now you can begin to understand the challenge of landing an aircraft. But don't worry, it gets better with practice. And the next lesson will give you the practice you need.

In this lesson, we covered the following topics:

Slow flight at 50-percent throttle
Slow flight and turns at 30-percent throttle
Stalls at 30-percent throttle
Power-off stalls in landing configuration
Spot view zoom controls

Land Ho! Landing the Aircraft

Sometimes a friend will point at a sleek-looking aircraft, such as a Learjet or a Cessna Citation, and ask whether I could fly it. My common reply is "Yeah, I could fly that!" And then I add, "Landing would be a big problem, but I could fly it all right." The point is, flying an airplane is easy; it's landing that's difficult. And every aircraft is different. Landing the Cessna 182 is not the same as landing the Learjet. But the concepts and procedures are similar, as you will see.

In our first approach to landing an aircraft (pardon the pun), we will apply the concepts that we learned in slow flight and stalls to our landings. This next lesson will basically be a slow flight ending in a stall except that this time the stall will happen just barely above the runway at Meigs Field. This situation is one that you can use over and over until you become comfortable with landings. Once you get the hang of it, I'm sure that you will be able to go on to bigger challenges, such as landing the Learjet. So let's get started with the lesson.

Flight Plan: This flight will start 5500 feet in the air on a northerly approach into Meigs Field. We'll use the Cessna 182, and we'll be in a power-off glide configuration. The purpose of this lesson is to demonstrate how landing an aircraft "feels." The only instruments we'll use will be the airspeed indicator, the altimeter, and the vertical speed indicator.

✝✝✝

Be sure to pause Flight Simulator before you enter the flight settings for this flight.

Flight Settings

Location	Approach to Meigs Field
Aircraft	Cessna Skylane RG R182
North/South Latitude	N041°46'22.3118
East/West Longitude	W087°36'26.0501
Altitude	5500 feet
Heading	000.82°
Auto Coordination	On
Season	Spring
Time of Day	9:00 a.m.
Realism and Reliability	
Prop Advance	Fixed Pitch

Enter the Flight Settings

1. Press P to pause Flight Simulator.
2. Open the Options menu, and select Aircraft.
3. Select the Cessna Skylane RG R182, and click OK.
4. Open the World menu, and select Set Exact Location.
5. In the North/South Lat. box, type **N041*46'22.3118**, and press Enter.
6. In the East/West Lon. box, type **W087*36'26.0501**, and press Enter.
7. In the Altitude box, type **5500**, and press Enter.
8. In the Heading box, type **000.82**, and press Enter.
9. Click OK to apply these settings.
10. Open the World menu and be sure Slew is not checked.
11. Open the Sim menu and be sure there is a check mark to the left of Auto Coordination. If not, click on the selection or press A.
12. Open the Sim menu again, and select Realism and Reliability.
13. Set the Prop Advance to Fixed Pitch, and click OK.
14. Open the Views menu, and select Set Spot Plane.
15. In the View Direction box, move the dot behind the airplane with your mouse or the arrow keys, and click OK.
16. Open the World menu again, and select Set Time and Season.
17. Set the Season to Spring.
18. Select Set Exact Time, type **9** in the Hours box, and press Enter. Type **00** in the Minutes box, and press Enter.
19. Click OK to apply these settings.
20. Open the Options menu, select Save Situation, type **CH2FLT3** in the Situation Title box, and then press Enter.
21. Click OK to save this situation.

Wing Tip:

Most people would say that a perfect landing is one in which the passengers are unable to determine when the aircraft has actually landed. But when I am flying, I define a perfect landing as one in which I manage to get all three wheels on the runway without damaging the aircraft or the composure of my passenger. So don't worry about trying to plant the aircraft exactly on the numbers or without a bounce or two. Just concentrate on getting it on the runway in one piece. The smoothness and technique will come later.

Approach to Meigs Field

Before you unpause the simulator, let me explain a little about what is going to happen. This landing will be a power-off glide to the airport. The engine will be on, but the throttle will be set at 0-percent power. Also, because Flight Simulator does not allow you to preconfigure throttle settings, we will have to start this flight at 0 airspeed. Consequently, the aircraft will make a small dive at the start of the flight until it gains enough airspeed to glide to the airport.

During the glide, we'll apply carburetor heat and lower the gear and the flaps to achieve a slow-flight configuration. We might need to make some minor adjustments to the heading in order to line up with the runway, but doing this will be good experience for the lesson on pattern flying. Our glide, if done correctly, should place the aircraft just on the numbers at the south end of the runway, the indication of a perfect approach and landing.

Let's get started by unpausing the simulator. Let the nose of the aircraft drop below the horizon. The airspeed will build back up, which will cause the aircraft to climb. Level off at around 4500 to 4700 feet instead of letting it climb, because you will need this airspeed to help you control the airplane as you start your glide into

Scenery

Meigs Field. As soon as the aircraft is level but descending, apply the carburetor heat. Airspeed should be around 100 knots.

Start the Approach

1. Press P to unpause the simulator.
2. Let the aircraft dive, and as airspeed increases, do not let the aircraft climb above 4700 feet.
3. Press H to turn on the carburetor heat.
4. Level off for a descent rate of around 100 knots.

Apply 10° of flaps. This will slow the aircraft down, increasing the rate of descent. Maintain a constant heading of 000°. As the nose tries to dip, keep enough back pressure on the yoke to maintain an airspeed of around 90 to 100 knots. As we pass through 4000 feet, increase the flaps to 30°. The airspeed will decrease to around 80 knots. When you increase the flaps, the nose of the aircraft will pitch up a little, due to the increased lift of the wing. Adjust for this by keeping the nose in the proper attitude for the 80-knot airspeed and approach to the runway. Remember, we are controlling the airspeed with the pitch of the aircraft, not with the throttle. Notice the visual position of the horizon in relation to the top of the instrument panel.

Chapter 2: Pilot Training and Private Lessons

*Aeronautically Speaking:
It is very important to apply the carburetor heat during landing approaches because doing so keeps ice from building up in the air intake of the carburetor. Even on hot days—and especially on humid days—ice can form in the carburetor air intake because the compressed air flow in the intake cools the air to below-freezing temperatures. The use of power is usually not critical in a gliding approach to an airport. However, if for some reason you are told by the tower to go around, or if you need additional power to adjust for the wind, the last few seconds above the runway would be the wrong time to find out that there is ice in the carburetor and that the engine does not want to power up.*

Add Flaps

1. Apply 10° of flaps by pressing F6.
2. Adjust yoke pressure to maintain an airspeed of 90 to 100 knots.
3. At 4000 feet, press F7 to apply 30° of flaps.
4. Adjust yoke pressure to maintain an airspeed of 80 knots.

As the aircraft passes through 3000 feet, apply full flaps. Pay close attention to the descent rate on the vertical speed indicator, which should register somewhere around 1200 feet per minute. The airspeed should now be around 70 to 75 knots. You will need to adjust the heading a little to compensate for the location of the runway. Aim the nose of the aircraft a little to the right of the runway as we pass through 2500 feet. When the runway appears to be pointed directly at the aircraft, bank slightly to the left to line the aircraft up with the runway. You might need to make further adjustments as we get closer to the runway.

Apply Full Flaps

1. At 3000 feet, apply full flaps by pressing F8.
2. Adjust yoke pressure to maintain an airspeed of 70 to 75 knots.
3. Compensate as needed for your heading to the runway.

Keep an eye on the airspeed indicator as we pass through 1500 feet. Remember, the altitude of the runway is at 600 feet. As we pass through 1000 feet, mentally prepare to *flare* the aircraft—that is, to level it off just above the runway. Begin to flare at around 800 feet. As we pass through 700 feet, bring the aircraft to straight and level flight at about 60 knots of airspeed. As the nose begins to point toward the horizon, gently apply more back pressure to the yoke, bringing the airspeed down to 50 knots just above the runway. The back wheels will touch down first, and then, as the airspeed decreases, the nose wheel will touch down. After all three wheels are down and the aircraft has slowed to 40 knots, apply the brakes.

Land the Aircraft

1. At 1000 feet, mentally prepare to flare the aircraft.
2. Begin the flare at 800 feet by applying gentle back pressure on the yoke.
3. At 700 feet, increase back pressure on the yoke to reduce the airspeed to 60 knots.
4. Just before touching the runway, stall the aircraft at 50 knots.
5. Maintain back pressure on the yoke after the main wheels touch.
6. Apply the brakes after the aircraft has slowed to 40 knots.
7. Pat yourself on the back.

Congratulations! You have just successfully landed the aircraft. If you had difficulty or were not able to land, keep trying. If you continue to have difficulty, you might want to have someone read the instructions to you as you go through this lesson. That way, you won't be distracted by reading instructions while you are trying to land.

As you get comfortable landing using the power-off glide approach, you might want to try some variations. Here are some suggested variations to try with the same situation:

Decrease the starting altitude to 4000 feet, and use the throttle to decrease your descent rate.
Increase the altitude to 6500 feet, and use the Learjet at 25-percent throttle.
Try the approach without using any flaps.
Change the exact time to 2200 hours (10:00 p.m.).
Fly the entire landing from Spot view behind the aircraft.

Pattern Flying

The next step in becoming a Flight Simulator private pilot is learning how to "fly the pattern." The *pattern* is a defined traffic route around an airport that is used for flying an approach to the airport. If an aircraft wanting to land at the airport is not on a straight-in approach, it needs to enter the pattern. The diagram should help you visualize the traffic pattern and the procedure for flying the pattern at Meigs Field.

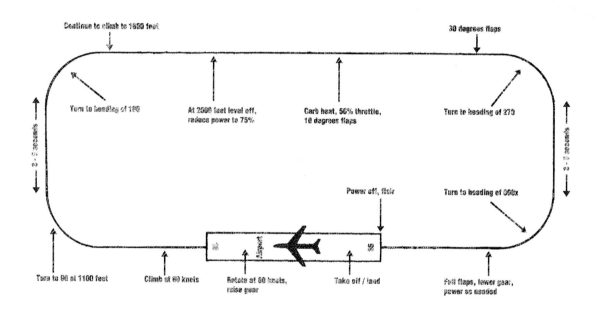

The air traffic pattern for Meigs Field in Flight Simulator 5. Although this diagram is related to real-world procedures, the actual air traffic procedures used at various airports may be different.

Flight Plan: This flight will be in the Cessna 182 at Meigs Field in Chicago. The flight will be local and will take place within the traffic pattern. We'll do three take-offs and landings to help you become familiar with the procedures involved with air traffic pattern flight.

✧ ✧ ✧

Flight Settings

Location	Meigs Field
Aircraft	Cessna Skylane RG R182
North/South Latitude	N041°51'13.3984
East/West Longitude	W087°36'27.2416
Altitude	594 feet
Heading	000.83°
Auto Coordination	On
Season	Spring
Time of Day	9:00 a.m.
Realism and Reliability	
Prop Advance	Fixed Pitch

Enter the Flight Settings

1. Open the Options menu, and select Situations.
2. Select Meigs Takeoff Runway 36, and click OK. (Note that if you have modified the Meigs Takeoff situation, you may have to enter the exact coordinates, altitude, and heading as described in the Flight Settings above.)
3. Open the Sim menu, and be sure there is a check mark to the left of the Auto Coordination selection. If not, click on the selection or press A.
4. Open the Sim menu again, and select Realism and Reliability.
5. Set Prop Advance to Fixed Pitch, and click OK.
6. Open the View menu, and select Set Spot Plane.
7. Move the dot in the View Direction box to the left of the airplane with the mouse or your arrow keys.
8. Set Transition to Fast.
9. Click OK to apply these settings.
10. Open the World menu again, and select Set Time and Season.
11. Set Season to Spring.
12. Select Set Exact Time, type **9** in the Hours box, and press Enter. Type **00** in the Minutes box, and press Enter.
13. Click OK to apply these settings.

The Takeoff Leg

Flying in the pattern can keep you very busy, so it's probably a good idea to read through the flight before you actually fly the lesson. Reviewing the diagram before the flight should also be helpful.

When you are ready to go, just apply full throttle. Keep the aircraft centered down the runway and let the airspeed build up. As the airspeed reaches 75 knots, start pulling back on the yoke. The aircraft will rotate at 75 to 80 knots. Drop the nose a little to keep the airspeed at 80 knots, and raise the gear. Keep the climb speed at 80 knots. Maintain this attitude until we reach 500 feet AGL (above ground level), or 1100 feet ASL (above sea level).

Take Off

1. Hold down keypad 9 to apply full throttle.
2. Pull back on the yoke at 75 knots.
3. After liftoff, apply pressure on the yoke to drop the nose a bit, and press G to raise the gear.
4. Maintain a constant airspeed of 80 knots during the climb.

The Crosswind Leg

When we reach 1100 feet ASL, make a standard-rate right turn to a heading of 90°. Continue to climb at 80 knots during the turn. After reaching a heading of 90°, level off and continue the climb at 80 knots for the next five seconds, before we make our next turn.

Complete the Crosswind Leg

1. At 1100 feet ASL, make a standard-rate turn to 90°.
2. At 90°, continue to climb at 80 knots for the next five seconds.

The Downwind Leg

Now make a standard-rate right turn to 180°. Continue to climb at 80 knots during the turn. Shortly after reaching a heading of 180°, you should reach an altitude of 2000 feet. At 2000 feet, lower the nose for straight and level flight. Use the elevator trim to reduce the pressure on the yoke, and reduce the throttle to 75-percent power. Switch to Spot view from the left of the aircraft, and watch for the aircraft to pass by the south end of the runway. As soon as it passes the end of the runway, switch back to Cockpit view, apply carburetor heat, reduce the throttle to 50 percent and add 10° of flaps. The airspeed should be around 90 knots.

Chapter 2: Pilot Training and Private Lessons

Again, switch to a Spot view, this time from the left front of the aircraft. Watch for the tail of the aircraft to pass by the end of the runway. At this point, lower the flaps to 30° and reduce the throttle to 30 percent. The airspeed should now be around 80 knots.

Complete the Downwind Leg

1. Make a standard-rate right turn to a compass heading of 180°, and continue to climb at 80 knots.
2. At an altitude of 2000 feet and a heading of 180°, lower the nose for level flight and press keypad 3 several times to reduce the throttle to 75-percent power.
3. Use keypad 7 and keypad 1 to trim the aircraft for straight and level flight.
4. Press S twice to switch to Spot view.
5. As the aircraft passes the south end of the runway, press S to switch back to Cockpit view.
6. Press keypad 3 several times to reduce the throttle to 50-percent power.
7. Press F6 to apply 10° of flaps.
8. Press S twice to switch to Spot view.
9. Press Shift-keypad 7 to select the left front Spot view of the aircraft.
10. When the aircraft's tail passes the south end of the runway, press S to switch back to Cockpit view.
11. Press F7 to apply 30° of flaps.
12. Use keypad 3 to reduce the throttle to 30-percent power.
13. Use keypad 7 and keypad 1 to trim the aircraft for a descent at 80 knots.

The Base Leg

Now we need to make a standard-rate right turn to a heading of 270°. We should be slowly descending now, and our airspeed should be around 80 knots. Continue this straight descent for the next five seconds.

Complete the Base Leg

1. Make a standard-rate right turn to a heading of 270°.
2. At 270°, continue a straight descent for the next five seconds.

The Final Approach

Okay, so far so good! Make a standard-rate right turn to a heading of 000°. As we come out of the turn, you will need to make whatever adjustments you think are necessary for lining up with the runway. Lower the gear and lower the flaps to full. This will drop our airspeed to around 65 to 70 knots. This part of the landing is a matter of getting used to how the aircraft is positioned in relation to the runway. If you feel that the glide path is too low, you should add some more power. If the glide path is too high, reduce the throttle a little. As in the lesson on the straight-in approach, it is not necessary to plant the aircraft right on the numbers at the end of the runway. If you need a little extra runway, go ahead and use it.

As the aircraft passes over the end of the runway, reduce the throttle to 0-percent power. Flare the aircraft just above the runway, and try to have the airspeed down to 50 to 55 knots when the wheels touch down. As soon as the nose wheel touches down, turn off the carburetor heat, raise the flaps to full up, and apply full throttle for another flight around the pattern. This is called a *touch-and-go*. Repeat this procedure for two more landings. On the last landing, come to a full stop.

Make the Final Approach and Land

1. Make a standard-rate right turn to a heading of 000°.
2. Adjust the heading as needed to line up with the runway.
3. Press G to lower the gear.
4. Press F8 to lower the flaps to 40°.
5. Press keypad 3 and keypad 9 to adjust the throttle as needed.
6. When you pass over the end of the runway, hold down keypad 3 to reduce the throttle to 0 percent.
7. Pull back on the yoke to flare the aircraft just above the runway.
8. Try to touch down at 50 to 55 knots.
9. Press H to turn off the carburetor heat.
10. Raise the flaps to full up by pressing F5 .
11. Hold down keypad 9 to apply full throttle for takeoff.
12. Repeat this lesson and make two more landings the same as the first one. On the third landing, press . (period) at 40 knots and come to a full stop.

That was a busy lesson, but it provides good training in controlling the aircraft. Once you can land the aircraft on your own, your confidence should greatly increase.

In this lesson we covered the following topics:

Takeoffs
Standard-rate turns
Slow flight
Climbs and descents
Landings
Air traffic pattern procedures

Time to Solo

Now that you are comfortable with flying and landing the aircraft, it's time to turn you loose. That's right, it's time for you to solo. This is basically repeating the last lesson by yourself. The only difference is that after you land, you will need to retract the flaps, turn off the carburetor heat, apply full throttle, and take off again. The third landing should come to a full stop. Use the diagram for the pattern flying to help you remember the procedures for the pattern.

Chapter **3**

Weather

"Did you get in touch with FSS?" the instructor asks the student.

"Yes, and they think that if we change our altitude to 7000 feet, we can get over the cloud layer."

"Well, I guess that will be okay. What is the rest of the weather advisory for our trip?"

"Fifteen miles visibility, winds aloft are 15 knots out of 30°, surface winds are 5 knots out of 40°, clouds are 2/8 coverage with base at 3000 and tops at 6000, and barometric pressure is 29.92."

"So flying at 7000 feet will indeed keep us 1000 feet above the clouds. Are you comfortable with that?"

"Sure! It gives me a chance to fly on top for awhile and stay VFR."

"Good! Go ahead and preflight 287, and I'll file the flight plan for us."

The Importance of Weather

Although it might be more enjoyable to fly in calm, clear conditions, it is more realistic to confront the ever-changing weather conditions that occur in real life. It is not uncommon to depart from the airport in completely clear weather only to find a storm front blocking the path to your destination airport. Being able to identify how the weather is changing can mean the difference between life (the fun of flying) and death (having to restart the situation).

Even though weather is often unpredictable and complex, understanding some basic factors can help any pilot determine or predict the weather that results from them. In Flight Simulator, not only can you predict the weather, you can create it. But before we create some of our own weather, let's learn about clouds, winds, temperature, and barometric pressure.

Clouds

Clouds are the most visible indicator of what the weather is like. A single glance at the clouds often enables a pilot to make a "go" or "no go" decision. Instrument rated pilots might not be as concerned, because they are trained to rely on instruments, but a VFR pilot pays strict attention to cloud formations and does whatever is possible to stay away from them.

In Flight Simulator, you can simulate various types of cloud coverage. The following table shows some of the settings that could be used for various types of cloud coverage:

Cloud Type	Description	Settings
Altostratus	Flattened and dense. Similar to stratus.	Tops: 20,000 feet Base: 6000 feet Coverage: 1/8 to Overcast
Cirrus	Wispy and flattened, with pointy edges.	Tops: 20,000 + feet Base: 18,000 + feet Coverage: 2/8 to 6/8
Cumulonimbus	Thick and puffy, solid white on top and dark on bottom. Produces thunderstorms.	Tops: 50,000 + feet Base: Ground + Coverage: 4/8 to Overcast
Cumulus	Fluffy, solid, "cottony," smaller than cumulonimbus.	Tops: 6000 – feet Base: Ground + Coverage: 4/8 to Overcast
Stratus	Flattened, elongated, and dense.	Tops: 6000 feet Base: Ground + Coverage: 4/8 to Overcast

Aeronautically Speaking: According to FAA regulations, a pilot licensed for VFR (Visual Flight Rules) must stay at least 1000 feet above, 500 feet below, or 2000 feet horizontally clear of a cloud formation. Failure to adhere to this rule can result in a suspension of all flying privileges. This rule is an important safety precaution because a VFR pilot does not have the training necessary to remain in full control of the aircraft without outside visual clues for navigation. In the clouds, a VFR pilot can become disoriented and lose control.

In Flight Simulator, you can create two cloud layers and one thunderstorm layer. For example, you could set up a layer of stratus clouds, a higher layer of altostratus clouds, and some scattered thunderstorms. Now that would make for an interesting flight!

At dawn and dusk, the clouds can turn pink or red against a dark blue sky. The change in the color has to do with how light from the sun passes through the atmosphere. Because the light enters horizontally instead of from directly above, it must travel a longer distance through the atmosphere and thus pass through more particles in the atmosphere. These particles scatter the light waves, resulting in more red, orange, and yellow colors in the sky. You can see these color changes in Flight Simulator by setting the time of day to dawn, dusk, or the time around sunrise and sunset for the selected season.

If you really want to impress someone with the visual capabilities of Flight Simulator, just set up a flight over Chicago at sunset with the clouds set for 1/8 coverage, stars on, approach lights on, and ground scenery on. It's a pretty sight! In fact, for our first weather flight, we are going to start out on a brisk Chicago morning with just a few clouds in the sky, and then we'll see how clouds can affect visibility.

Flight Plan: We will be departing from Meigs Field in Chicago in the Cessna Skylane RG R182. Our flight will take us into the clouds so that we can experience reduced visibility. We will also change the various settings for cloud coverage and see how the changes affect our ability to see the ground.

+++

Flight Settings

Location	Meigs Field, Runway 36
Aircraft	Cessna Skylane RG R182
North/South Latitude	N041°51'13.3984
East/West Longitude	W087°36'27.2416
Altitude	594 feet
Heading	000.83°
Auto Coordination	On
Season	Winter
Time of Day	6:30 a.m.
Realism and Reliability	
Prop Advance	Fixed Pitch
Weather	
Clouds	
Base	5000 feet
Tops	6500 feet
Coverage	1/8, Scattered
Winds Aloft	
Type	Steady
Base	3000 feet
Tops	8000 feet
Speed	15 knots
Direction	30°
Turbulence	1
Temperature	
Altitude	3000 feet
Daytime Temperature	25°F
Day/Night	Yes
Barometer	
Pressure	29.92Hg
Drift	None

Enter the Flight Settings

1. Open the Options menu.
   Select Situations.
   Select Meigs Takeoff Runway 35.
   Click OK.
   (Note: If you have altered the default Meigs Takeoff Runway 36 situation,
   please refer to the flight settings for the exact location and aircraft for this flight.)

2. Open the World menu.
   Select Set Time and Season.
   Set Season to Winter.
   Select Set Exact Time.
   Type **6** in the Hours box, and press Enter.
   Type **30** in the Minutes box, and press Enter.
   Click OK.

3. Open the World menu.
   Select Weather.
   Select Edit in the Cloud Layers box.
   Type **5000** in the Base box, and press Enter.
   Type **6500** in the Tops box, and press Enter.
   Check to make sure that the Coverage box shows Scatter 1/8.
   Click OK.

4. Select Winds.
   Select Create.
   Select Steady in the Type box.
   Type **3000** in the Base box, and press Enter.
   Type **8000** in the Tops box, and press Enter.
   Type **15** in the Speed box, and press Enter.
   Type **30** in the Direction box, and press Enter.

5. Select Temp.
   Select and delete all temperature layers.
   Select Edit.
   Type **3000** in the Temperature Altitude box, and press Enter.
   Type **25** in the Daytime Temperature box, and press Enter.
   Click OK to return to the Weather dialog box.
   Be sure the Day/Night box is checked.
   Select Baro. and be sure that Pressure is 29.92 and that Drift is not selected.
   Click OK again to apply these settings.

6. Open the Sim menu, and be sure that Auto Coordination is checked.
   Select Realism and Reliability.
   Select Fixed Pitch in the Prop Advance box.
   Click OK.

7. Open the Options menu.
   Select Save Situation.
   Type **CH3FLT1** in the Situation Title box, and press Enter.
   Feel free to type your own description of this situation in the Description box, and then press Enter.
   When you are finished, click OK to save this situation.

Head in the Clouds

Now, doesn't that sky look beautiful? There is something to be said about early morning flights—just a touch of pink cloud in the sky, the horizon is starting to lighten up, and the sky is a wonderful shade of blue. The red glow of the instrument lights contrasts with the colors of the sky. Meigs Field is familiar enough, but you might need some background on the weather over Chicago.

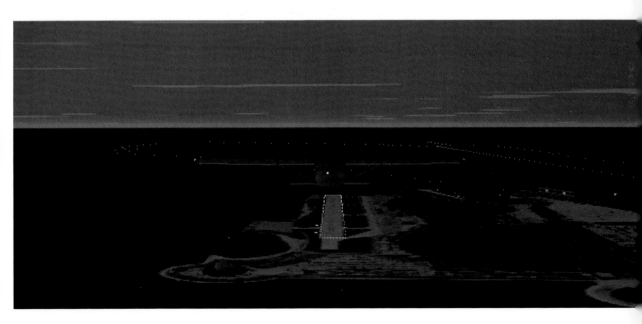

They don't call Chicago the windy city for nothing! In the summertime, you could take off into a clear sky and find yourself facing a storm front off Lake Michigan within an hour. The winds can really make it interesting when you're trying to make a sluggish airplane climb in 95° temperatures with 90-percent humidity. (The high temperature and humidity affect the ability of the aircraft to climb, which we'll discuss later.) But fortunately, for this flight we'll be flying in the winter at 6:30 in the morning, and the weather should be pretty calm.

All set to go now? Good. Because it is still a little dark, it would be a good idea to turn on the lights while we depart. Turn on both the landing lights and the instrument lights. Go ahead and start the takeoff, and then rotate at 80 knots. Keep the plane centered on the runway. Once you get the plane up off the ground, drop the nose just a touch to allow the airspeed to build up. Then raise the gear. Trim the plane to climb at 95 knots, which should give you a climb rate of around 1000 feet per second. As we pass 1000 feet, look behind us for a nice view of Meigs Field in the dawn's early light.

Continue your climb with the heading of 000°. At 2000 feet, look for other traffic out the left and front of the aircraft, and then begin a standard-rate left turn to a heading of 180°. Begin to level off the turn at 10° before the desired heading of 180°, but continue to climb.

Take Off and Turn While Climbing

1. If the instrument and landing lights are not on, press L to turn them on.
2. Press the . (period) key to release the parking brakes.
3. Apply full throttle for takeoff.
4. Rotate at 80 knots and, when airborne, press G to raise the gear.
5. Use keypad 7 and keypad 1 to trim the aircraft for a climb speed of 95 knots.
6. Press S twice to change to Spot view at 1000 feet, and then press Shift-keypad 8 to take a look at the airport from in front of the aircraft as you ascend.
7. Press S to return to Cockpit view.
8. Make a standard-rate left turn to 180° at 2000 feet while continuing your climb.

Cloud Settings

You'll notice that with a cloud setting of 1/8, there are just enough clouds to make the sky look pretty but not enough to cause any problems for a VFR pilot. Now let's see how the different cloud settings affect the amount of clouds in the sky. As you pass through 2500 feet, change the cloud settings for a coverage of 2/8. (Go ahead and use the Pause key if you get a little fumble-fingered at the keyboard or need to look up some key commands in the *Flight Simulator Pilot's Handbook*.)

Change the Cloud Coverage to 2/8

1. Open the World menu, and select Weather.
2. Select Clouds, and then select Edit.
3. Change the coverage to Scatter 2/8, and then click OK.

With a cloud coverage of 2/8, we can see more clouds but still not enough to hinder visual flight. As we pass through 3500 feet, select a new cloud coverage setting of 4/8. Now we'll see a lot of continuous coverage, with an equal amount of blue sky. With the deviation setting of 500, we are able to create a more realistic cloud arrangement in the sky. The deviation number is used by Flight Simulator for generating a random pattern for the frequency and appearance of the clouds. As we pass through 4000 feet, go ahead and change the cloud settings to 7/8, and notice that the coverage is almost 100 percent. This type of cloud coverage would be very difficult for a VFR pilot to fly through because the patches of clear sky are few and almost too small to fit an aircraft through without breaking the 2000-foot clearance rule.

Change the Cloud Coverage Settings

1. At 3500 feet, open the Weather menu and change cloud coverage to Scatter 4/8, type **500** in the Deviation box, and press Enter.
2. Click OK twice to apply these settings.
3. At 4000 feet, change cloud coverage to Broken 7/8 using the same steps as above.

Into the Clouds

Continue the climb into the base of the clouds. Notice how the clouds start to affect your view through the windshield. Because many of the clouds are small puffs hanging under a solid base, they will affect your view out the cockpit windshield by causing it to become "fogged" and then clear again. It's kind of like driving in a car through patches of ground fog.

Take a quick look for other traffic just before you reach 5500 feet. It's always good to make sure that no other aircraft are entering the clouds at a speed and angle that would place them in your path. After you are in the clouds, it is impossible to see any other traffic until it is too late.

When you enter the clouds above 5500 feet, you'll notice that you will not be able to see anything out the windshield. Although you haven't learned a lot about instruments yet, it is a good idea to keep an eye on the attitude indicator and the vertical speed indicator. These will give you a good visual representation of the attitude of the aircraft. Regardless of what you may think the aircraft is doing in relation to the ground, always rely on the instruments to tell you what is really going on. We'll discuss more of the instruments in the near future, when we begin instrument training.

As we pass through 6000 feet, we will come out of the solid cloud mass and enter into patchy clouds again. We should still be climbing at 95 knots. At 6500 feet, we will break out on top of the clouds. Notice that there are no more clouds above us; the cloud cover is now below. This is why it is important to learn to use the instruments for navigation. With the ground obscured from view by the clouds, there are no visual reference points for navigation. Go ahead and level the nose of the aircraft at 7000 feet, and then bring the power back to 75 percent. Remember to trim the aircraft for the new attitude.

Level Off

1. Level off at 7000 feet.
2. Press keypad 3 to reduce power to around 75 percent. The throttle indicator should be at about one-fourth the distance from the top.
3. Use keypad 7 and keypad 1 to trim the aircraft for straight and level flight.

Out of the Clouds

This would be a good time to check out the various views. Feel free to switch to Spot view and look around at the aircraft and at the clouds below. If something looks special, you can always take a flight photograph as well. When you have finished, return to Cockpit view, clear the area off to the left (look for traffic), and then make a standard-rate left turn to a heading of 000°. Maintain your altitude throughout the turn. Again, remember to start coming out of the turn 10° before the target heading. When we reach the desired heading, you can begin a standard approach into Meigs Field for landing.

Examine the Views and Turn to 000°

1. Press S twice to switch to Spot view, and use Shift and the keypad keys to look around. Use the Flight Photograph command from the Views menu if desired.
2. Press S to switch back to Cockpit view.
3. Make a standard-rate left turn to a heading of 000°.
4. Begin a standard approach into Meigs Field, or press X to use the Land Me feature.

Okay, you're on your own now. Remember what was covered in the earlier lessons about making the approach, using flaps, and especially using the throttle to control the descent rate. If the aircraft seems to drift as you head back toward the airport, it is because you have a 15 mph wind. After you descend below 3000 feet, this drift should disappear. If you have difficulty, use the Pause key while you refer to the earlier lessons. Or you can always revert to using the Land Me feature (press X).

Everyone Knows It's Windy

Unlike clouds, wind is not directly visible, but it plays a significant role in flying conditions nonetheless. Not many Flight Simulator pilots use the wind settings, because doing so alters their ability to take off and land smoothly. Landing with a 10-knot crosswind is a lot of work, but it is also realistic.

Many factors affect the formation and activity of wind. The movement of the earth, temperature, air pressure, ocean currents, and—as a book on chaos theory suggests—even the beating wings of a butterfly can affect the wind. Pilots should always check with the local flight service station (FSS) or a similar aviation-related weather service for the recent wind advisories and forecasts. It might be calm on the ground, but once you get into the air it can be a different story. Nice and calm at 500 feet can be 15 knots of headwind with turbulence at 3000 feet.

Because of changes in temperature and reduced pressures at higher altitudes, wind speed can increase greatly as you climb higher. Heavier aircraft can even take advantage of jet streams at 45,000 feet (unrealistic for a Cessna R182 but very realistic for a Learjet). A typical jet stream over the northern United States during the winter would average 80 to 125 mph and would be moving from west to east.

You can create your own wind layers and jet streams with the Weather command in Flight Simulator. Up to three layers of winds aloft (above ground and at altitude) and one surface wind layer can be created. A typical realistic setting for winds could be:

Wind Layer	Settings
Layer 1— Aloft	Type: Steady Base: 30,000 feet Tops: 50,000 feet Speed: 135 knots Direction: 120° Turbulence: 1
Layer 2—Aloft	Type: Steady Base: 15,000 feet Tops: 30,000 feet Speed: 60 knots Direction: 90° Turbulence: 1.5–3
Layer 3—Aloft	Type: Gusty Base: 3000 feet Tops: 15,000 feet Speed: 10–20 knots Direction: 80° Turbulence: 1.5–3.5 (more, if you're sadistic)
Layer 4—Surface	Type: Gusty Depth: 3000 feet Speed: 5–10 knots Direction: 90° Turbulence: 1–1.5

Chapter 3: Weather

Helicopters R Us

How about a little trick flying? We'll perform this feat with your plane parked in the middle of Meigs Field, so you will need to change only a few settings for this next flight. Before changing these settings, pause Flight Simulator (press P).

Flight Plan: In the Cessna R182 from Meigs Field, we will attempt to make a vertical takeoff in an 80-knot headwind.

This flight begins with most of the same settings as the last flight, so there is no need to key in the location and configuration settings again. Just add the following settings:

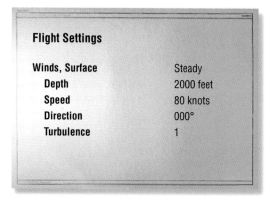

Enter the Flight Settings

1. Press P to pause the simulator.

2. Open the Options menu.
   Select Situations.
   Select Meigs Takeoff Runway 36.

3. Open the World menu.
   Select Set Exact Location.
   Type **N041*51'37.4600** in the North/South Lat. box, and press Enter.

4. Open the World menu again.
   Select Weather, Winds, and Create.
   Select Surface Wind.
   Be sure the Type box shows Steady, the Direction box shows 000, and the Turbulence gauge is set to 1.
   Type **2000** in the Depth box, and press Enter.
   Type **80** in the Speed box, and press Enter.
   Click OK twice.

5. Open the Options menu.
   Select Save Situation.
   Type **CH3FLT2** in the Situation Title box and press Enter. Feel free to type your own description of this situation in the Description box, and then press Enter.
   Click OK.

Wing Tip:
You can also try this flight while watching your vertical takeoff from the Meigs Tower. Simply press S to switch to Tower view before starting your "helicopter" flight! You may even want to click on your zoom indicator to get different distance views from the tower. Clicking on the zoom number reduces the amount of zoom, and clicking on the brackets on the right of the zoom number increases the amount of zoom.

Because we now have a headwind of 80 knots, the aircraft will want to roll backward. As soon as we unpause, give the aircraft enough throttle to keep the speed at 80 knots. It's going to be a little bit tricky, but try to adjust the throttle so that the aircraft is stationary and is not being pushed backward.

Throttle Up

1. Press P to unpause the simulator.
2. Press . (period) to release the parking brakes (if set).
3. Use keypad 9 to apply enough throttle to remain stationary.

We Have Liftoff

Now gently pull back on the yoke just enough to get the plane to start climbing, but be sure that the airspeed stays over 65 knots. Go ahead and climb to 800 feet. Notice how the runway does not appear to be moving under the plane. Once we reach 800 feet, drop the nose slightly and glide in for a normal landing on the runway. Remember to maintain a little bit of throttle to keep from being pushed backward. After landing, be sure to pat yourself on the back for flying the first VTOL (Vertical Take Off and Landing) Cessna R182!

Chapter 3: Weather

Wing Tip:
In the real world, temperature has a great effect on the formation of winds, pressure, and clouds. However, in Flight Simulator, temperature affects only airspeed. You can change the temperature with the Weather command on the World menu.

Perform a Vertical Takeoff

1. Gently apply back pressure on the yoke for takeoff.
2. Climb to 800 feet.
3. Lower the nose and glide in for a normal landing.
4. Use keypad 9 and keypad 3 to adjust the throttle, maintaining an airspeed of 80 knots.

Actually, no sane pilot would attempt to take off in 80-knot winds, but being able to simulate the critical or seemingly impossible situation is what makes Flight Simulator so much fun. To understand why a pilot would not even consider flying in such conditions, go through the last flight again. But this time, apply full power and fly a standard pattern and approach. To make it more realistic, be sure to change the winds to gusty instead of steady. If you make it back down in one piece on the correct end of the runway, go buy yourself a drink. You deserve it!

Too Hot to Handle

Temperature can also have a great effect on how your aircraft performs. An aircraft wing creates lift by forcing the air molecules over the top of the wing, thus creating a low-pressure area that lifts the wing up. The more molecules that are pushed over the wing, the more lift the wing is able to generate. When the air temperature is 40°F, there are a lot more molecules in a cubic foot of air than there are when the temperature is 80°F. Thus, on a hot summer day an aircraft needs a lot more runway to take off than it does on a cold winter day. Humidity also lessens the effect of the lift. The higher the humidity, the less lift the aircraft has.

Temperature also affects the airspeed needed to keep the plane in the air. It can mean the difference between a stall speed of 50 knots when it is cool outside and a stall speed of 55 knots when the temperature is high, which in turn can mean the difference between making it to the runway during a landing or winding up in the cow pasture a hundred yards short.

Within Flight Simulator, you can define four different temperature layers. A typical temperature configuration for winter in North America could be as follows:

Temperature Layer	Settings
Layer 1	Altitude: 500 feet Temp: 10°F
Layer 2	Altitude: 10,000 feet Temp: −20°F
Layer 3	Altitude: 20,000 feet Temp: −44°F
Layer 4	Altitude: 30,000 feet Temp: −67°F

I Can't Stand the Pressure!

Air pressure is another invisible factor in the formation of weather and its effect on flight. Just as temperature changes the density of air molecules in the air, air pressure is the direct representation of the amount of force being exerted by the air molecules. This is usually measured in inches of mercury. The average air pressure setting at sea level is 29.92 inches.

This measurement is important for flight, because the aircraft's altimeter uses pressure to determine the altitude of the aircraft. Before taking off, it is very important to get the local pressure reading and use it to set the altimeter. If we were flying between two distant locations, it would be important to find out what the altimeter setting should be for the destination airport. A discrepancy of one-tenth of an inch

Wing Tip:
Just as with temperature, air pressure has a great effect on the formation of weather in real life. In Flight Simulator, barometric pressure affects only the altimeter setting. You can adjust the barometric pressure with the Weather command on the World menu. You can also add a realistic drifting of the altimeter by choosing the Drift check box.

Aeronautically Speaking:
A good pilot checks with each local FSS for the correct altimeter setting on a regular basis during a long flight, particularly at the destination airport. For example, between Kansas City, Kansas and Denver, Colorado the ground appears to be level throughout the flight, but there is actually an increase of over 3000 feet in ground level. It would be very unwise for a pilot to fail to adjust the altimeter regularly during the flight.

Chapter 3: Weather

of mercury can cause an altimeter to display an error of 100 feet of altitude. If we were approaching an airport that was sitting 300 feet above mean sea level (MSL) and the altimeter indicated 100 feet less than it should, we could be in for major trouble, especially if we were flying in bad weather and had to rely completely on the instruments.

Weather Areas

It is fine to simulate a general weather pattern for an entire flight, but this is not typical of weather conditions that are found in real-world flying. In Flight Simulator 5, you can now create weather fronts and patterns for various areas. To demonstrate this capability, let's create a weather area and fly through it.

Flight Plan: Departure for this flight will be from John F. Kennedy International Airport in New York. We will be flying the Learjet and will take off in beautiful weather with only a few clouds. We'll fly out over New York City and will encounter a full overcast sky. From that point, you can do what you want with the flight.

✢✢✢

Adventures in Flight Simulator

Form 7233-1
04 R0072

Flight Settings

Aircraft	Learjet 35A
North/South Latitude	N040°37'20.6864
East/West Longitude	W073°47'06.9685
Altitude	17 feet
Heading	43.82°
Auto Coordination	On
Season	Summer
Time of Day	10:00 a.m.
Weather	
Weather Area	Global
Clouds	
Base	10,000 feet
Tops	11,000 feet
Coverage	Scattered 1/8
Weather Area	New York
Beginning Latitude	N041°00'00.0000
Beginning Longitude	W075°00'00.0000
Ending Latitude	N039°00'00.0000
Ending Longitude	W075°00'00.0000
Width	50 miles
Clouds	
Base	7000 feet
Tops	15,000 feet
Coverage	Overcast
Deviation	500 feet
Temperature	
Altitude	2401 feet [default]
Daytime Temperature	25°F

Enter the Flight Settings

1. Open the Options menu, and select Aircraft.
   Select Learjet 35A.

2. Open the World menu.
   Select Set Exact Location.
   Set North/South Lat. to **N040*37'20.6864** and East/West Lon. to **W073*47'06.9685**.
   Set Altitude to **17** and Heading to **43.82**.
   Click OK.

3. Open the World menu.
   Select Set Time and Season.
   Set Season to Summer.
   Select Set Exact Time.
   In the Hours box, type **10**, and press Enter.
   In the Minutes box, type **0**, and press Enter.
   Click OK.

4. Open the World menu.
   Select Weather.
   Select Edit in the Cloud Layers box.
   Type **10000** in the Base box, and press Enter.
   Type **11000** in the Tops box, and press Enter.
   Check to be sure that the Coverage box shows Scatter 1/8 .
   Click OK.

5. Select Add Area.
   Type **New York** in the Area Name box.
   Type **N041** in the Beginning Lat. box, and type **W075** in the Beginning Lon. box.
   Type **N039** in the Ending Lat. box, and type **W075** in the Ending Lon. box.
   Type **50** in the Width box, and then click OK.
   Select New York from the Weather Area menu.
   Select Clouds.
   Select Create in the Cloud Layers box.
   Select Overcast from the Coverage menu.
   Type **7000** in the Base box, **15000** in the Tops box, and **500** in the deviation box, and then click OK.
   Select Temp.
   Select Create. Make sure the Daytime Temperature is +025.
   Click OK twice to apply these settings.

6. Open the Sim menu and make sure Auto Coordination is checked.

7. Open the Options menu.
   Select Save Situation.
   Type **CH3FLT3** in the Situation Title box, and press Enter.
   Feel free to type your own description of this situation in the Description box, and then press Enter.
   Click OK to save this situation.

Creating a Weather Area

Before we begin our flight, you should know more about creating a weather area. At first, using the beginning latitude/longitude and the ending latitude/longitude can be a little confusing when you're trying to figure out where to place the weather area. The following illustration uses the settings that you entered for this flight and will help you to visualize the weather area:

Wing Tip:
This flight introduces the Learjet. The controls for the Learjet are more sensitive than the controls for the Cessna. Remember to use a gentle touch when making turns, descending, and climbing. During takeoff, due to the speed of the aircraft, it is very easy to place the Lear into a very steep climb attitude, resulting in a stall or a difficult-to-control situation. Just follow the instructions carefully and you'll enjoy the flight. More details about the Learjet will be introduced later in the book.

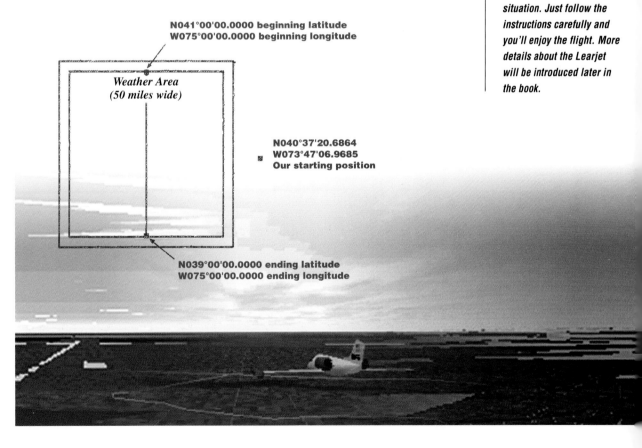

Chapter 3: Weather

Our starting location is N040°37'20.6864 by W073°47'06.9685. Because the weather area is to be placed off to the west of our starting location, the west longitude selected is greater than that of our current position. You'll notice that the actual position of the weather area is derived by drawing a line from the beginning and ending latitude/longitude positions. Then the size of the weather area is determined by the width parameter. Deviation varies the height of the cloud tops and base by a random number between 0 and the deviation factor you choose. This adds even more realism to weather areas in Flight Simulator.

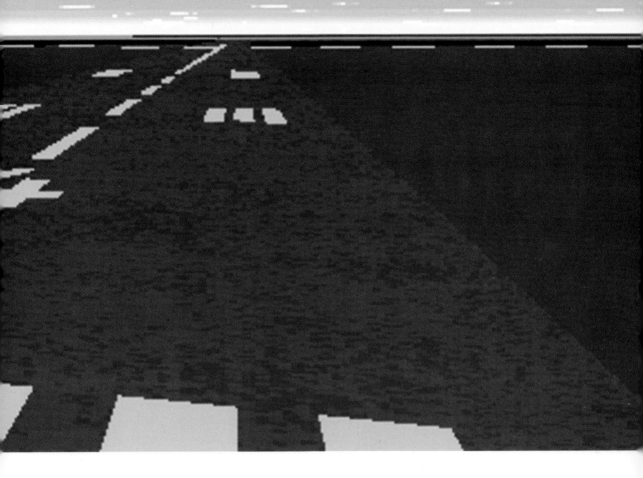

Clear-Weather Takeoff

Now that we know how to create all this weather, let's try it out. Unpause the simulator, and apply full throttle. Slowly pull back on the yoke at 130 knots. As soon as the Learjet lifts off the runway, you will need to gently lower the nose a little to keep the climb from being too steep. Raise the gear and then lower the nose until the horizon appears just below the top of the instrument panel. (You'll be able to see the horizon off to the left side of the instrument panel.) Keep climbing until we reach 5000 feet, and then level off.

Take Off

1. Press P to unpause the simulator, if necessary, and press . (period) to release the parking brakes if they are on.
2. Use keypad 9 to apply full throttle.
3. Gently pull back on the yoke at 130 knots and, after rotation, lower the nose a little to keep from climbing too steeply.
4. Press G to raise the gear, and lower the nose again until the horizon is just below the instrument panel.
5. Level off at 5000 feet.

Notice how the clouds look? Just a few clouds and lots of blue sky. Go ahead and reduce the throttle until the turbine speed gauge displays 80 (for 80 percent). Make a left turn to a heading of 270°. Try to start the turn before passing over the water below. Due to the speed of the Learjet, the turn will take a lot longer than it does in the Cessna. After coming out of the turn, you should be able to see New York City off to the left. Also notice that the clouds look a little thicker out over the horizon. Make another left turn until the nose of the aircraft is pointed over the top and slightly to the right of the buildings of New York.

Turn Toward New York City

1. Use keypad 3 to reduce the throttle until the turbine speed gauge displays 80.
2. Make a left turn to a heading of 270°. Notice the clouds in the distance.
3. Make another left turn to point the nose of the aircraft to the right of downtown New York City.

The Weather is Changing

After we pass over New York City, notice that the clouds are becoming thicker. After we fly into the clouds, go ahead and apply full throttle and climb to a new altitude of 17,000 feet. After reaching 17,000 feet, level off and reduce the throttle to 80 percent. Notice the complete cloud cover below? This is the weather area that we defined at the start of this flight. The change from clear to overcast weather was gradual, as it is in the real world. Now you know how to create and fly into a predefined weather area. If you want, you can continue to fly around or just keep the current heading, and soon you'll see the overcast clouds fade away as you fly out of the New York weather area.

In this chapter we covered the following topics:

Creating and changing clouds and how clouds affect flight
Creating and changing winds and how winds affect flight
The effect of temperature settings
The effect of barometric air pressure settings
Creating and changing weather areas

Now that you have a better understanding of the available weather within Flight Simulator 5, let's move on to learn the instrument skills you'll need for flying in the thick of it.

Chapter 4

Instrument Flight

The pilot scans the clouds below, trying to visualize where the airport might be. After changing the radio frequency, the pilot calls for assistance, "Martha's Tower, this is Cessna 22287, north of No Man's Land, with information Xray. Requesting ILS approach to runway 33."

"Roger, triple two eight seven. Squawk ident 2240."

"287 squawking 2240," the pilot repeats, changing the frequency on the transponder and pressing the squawk button.

"Cessna 22287, we have you on radar at 5000. Turn right heading 110 degrees, descend and maintain 2500."

"Roger, 287 descending to 2500, 110 degrees," the pilot responds with relief, knowing that the tower and the instruments will help guide the aircraft home.

Instrument School

Ready for some new challenges? If so, it's time to move into the ranks of the instrument flight rules (IFR) pilot. Visual flight rules require—not surprisingly—good visibility. For the visual flight rules (VFR) pilot, coastal fog or low cloud cover can mean a significant delay in getting off the ground. But the IFR pilot has no problem taking off and flying in such conditions. The difference is another level of training. The IFR pilot is skilled in reading and relying on the aircraft's instruments in situations in which visibility is poor.

The nice thing about Flight Simulator is that you can start flying right away and learn about IFR flight in the air. You can also pause the flight at various times while you learn more about related topics. So be prepared to use that P key on your keyboard even if it's only for time to wipe the sweat from your hands! But before you enter the flight settings, let's go over a few new settings that we will be using.

Using Noncoordinated Ailerons and Rudder

When you turn off the Auto Coordination command, you use separate controls for the ailerons and rudder, which more realistically simulates flying. Most other Flight Simulator pilots who have made the transition agree that after you get used to using separate controls, you'll find that controlling the aircraft is much easier. In fact, with a strong crosswind, it is almost impossible to land an aircraft without controlling the ailerons and rudder separately. You will still need to keep the movement of the ailerons and the rudder coordinated during turns, of course; you will simply be using two different controls to do this instead of one.

Using separate controls for the ailerons and the rudder can also come in handy when the aircraft is on a high approach to the runway. A good way to lose a lot of altitude but stay on target for the runway is to perform a slip. To do this, apply full rudder in one direction and apply opposite ailerons. It's a little awkward, but it works well for losing altitude fast.

Adjusting the Realism Settings

Until now we've used the easiest flight characteristics as we controlled the aircraft. Now let's change the characteristics—but only a little. Going from the easiest to the most realistic would be too drastic. We'll change the Flight Control Realism setting from 0 to 3, which will make the aircraft a little more sensitive to the controls.

Let's also turn on the following options to provide a more realistic flying environment:

- **Elevator Trim**   This option allows you to use the elevator trim to keep the aircraft in the desired attitude without having to keep constant pressure on the yoke.
- **Gyro Drift**   Because the directional gyro operates mechanically, it tends to drift from the true heading. With this option on, you will need to recalibrate regularly during a flight.
- **Airframe Damage from Stress**   When this option is on, you lose control if the airspeed and stress limitations of the aircraft are exceeded.
- **Engine Stops When Out of Fuel**   Ever have to walk to a gas station? When this option is on, instead of pulling over you'll have to find a place for an emergency landing.

Wing Tip:
If you are using the keyboard to control the aircraft, use the 0 key on the numeric keypad to move the rudder left and the + (plus) or Enter key on the numeric keypad (whichever is on the bottom right of the keypad) to move the rudder right. You can make a coordinated left turn by pressing keypad 4 and keypad 0 simultaneously. For a coordinated right turn, press keypad 6 and either keypad + (plus) or Enter simultaneously. If you keep the ball centered in the turn and bank indicator, your turn will be balanced and coordinated. If you are using a single joystick, you can still use the numeric keypad for rudder control. Pressing keypad 5 returns the rudder to its neutral position.

Chapter 4: Instrument Flight

- **Instrument Lights**   With this option, the instrument lights don't come on automatically; you'll have to click the switch when it gets dark.

- **Lights Burn Out**   When this option is on, the instrument lights don't automatically turn off in daylight. You'll have to remember to turn them off at the appropriate time, or some of them will burn out.

- **Fuel Tank Selector**   This option requires you to manually switch to the other fuel tank when the first one gets too low.

Hundred-Dollar Hamburgers

For our first IFR outing, let's take a $100 hamburger flight, so called because we'll fly from Santa Ana, California, to the island of Catalina, only 26 miles off the coast. The airport on Catalina has a nice little restaurant that actually serves a reasonably priced hamburger or even a buffalo burger. But in real-world aviation, by the time you buy fuel and pay the expense of the round-trip flight, the total cost is usually around $100!

Flight Plan: In the Cessna R182, we will depart at 11:30 a.m. from John Wayne Airport in Santa Ana, California, and fly via instruments to Catalina Island. A marine layer (fog) will still be hanging over the area, so we will encounter clouds with a ceiling (base) of 2500 feet. We will climb to an altitude of 4500 feet to get over the clouds and will use the Seal Beach very high frequency omnidirectional range (VOR) and the Santa Catalina VOR for navigation as well as a Los Angeles sectional chart.

✝✝✝

Flight Settings

Location	John Wayne Airport
Aircraft	Cessna Skylane RG R182
North/South Latitude	N033°40'53.0600
East/West Longitude	W117°51'54.8101
Altitude	57 feet
Heading	193.27°
Season	Summer
Time of Day	11:30 a.m.
Weather—Clouds	
Base	2500 feet
Tops	3500 feet
Coverage	Overcast
Auto Coordination	Off
Realism and Reliability	
Flight Control Realism	3
Elevator Trim	On
Gyro Drift	On
Airframe Damage from Stress	On
Engine Stops When Out of Fuel	On
Instrument Lights	On
Lights Burn Out	On
Fuel Tank Selector	On
Navigation Instruments	
NAV Radio 1	111.4
NAV Radio 2	115.7
COM Radio	126.0
ATC	Selected

Wing Tip:
Use a real Los Angeles sectional chart (the map) or the sectional chart included in the Flight Simulator Pilot's Handbook. *Then you can draw your own lines and make notes directly on the map to pinpoint your exact location during the flight. Placing the map on a clipboard and leaving it on your lap will add to the realism of the flight. Most pilots use clipboards that fasten to one of their legs or to the center of the yoke. You can buy sectional charts at almost any general aviation airport or flying school. (Who knows? Maybe you should sign up for some flying lessons while you're there!)*

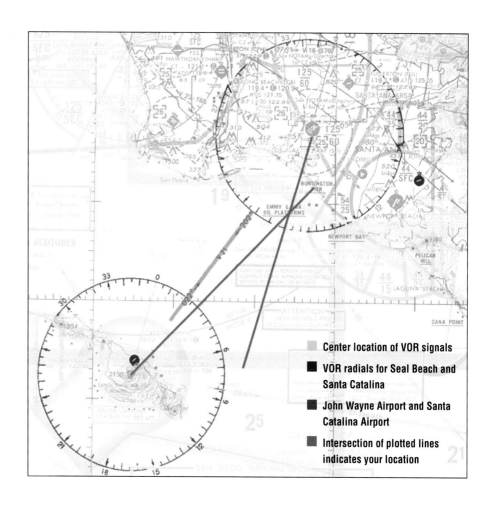

■ Center location of VOR signals

■ VOR radials for Seal Beach and Santa Catalina

■ John Wayne Airport and Santa Catalina Airport

■ Intersection of plotted lines indicates your location

Enter the Flight Settings

1. Open the Options menu.
   Select the Cessna Skylane RG R182 as your aircraft.
      (From now on, we will assume you know to click
      OK to apply settings when you are done with a dialog box.)

2. Open the World menu.
   Select Set Exact Location.
   Set North/South Lat. to **N033*40'53.0600** and East/West Lon. to **W117*51'54.8101**.
   Set Altitude to **57** and Heading to **193.27**.

3. Open the Sim menu, and be sure Auto Coordination is not selected.
   Select Realism and Reliability.
   Set Flight Control Realism to 3.
   Select Elevator Trim, Gyro Drift, Airframe Damage from Stress, Engine Stops
      When Out of Fuel, Instrument Lights, Lights Burn Out, and Fuel Tank Selector.

4. Open the World menu.
   Select Set Time and Season.
   Set Season to Summer.
   Set Exact Time to 11:30 a.m.

5. Open the World menu again, and select Weather.
   Edit Clouds to have a Base of **2500**, Tops of **3500**, and a Coverage of Overcast.

6. Open the Nav/Com menu.
   Select Navigation Radios.
   Set NAV 1 Frequency to 111.4 and NAV 2 Frequency to 115.7.

7. Open the Nav/Com menu again.
   Select Communication Radio.
   Set the COM 1 Frequency to 126.0.

8. Open the Nav/Com menu again.
   Select Air Traffic Control.

9. Open the Options menu.
   Select Save Situation.
   Save this situation as **CH4FLT1**.

Aeronautically Speaking:
ATC departure control is the aircraft traffic coordinator for aircraft departing from the airport. Departure control keeps track of aircraft wanting to depart and makes sure everyone gets a turn. Departure control also ensures that the spacing between the departing aircraft is enough to avoid midair collisions and allows time for the jet blast from a departing aircraft to dissipate before letting another aircraft take off.

Before we take off, let's look at the sectional chart to become familiar with the location of the airports and the VORs that we will be using.

Because we are flying IFR, our flight is going to require a lot of communication with the tower and with Los Angeles Center air traffic control (ATC). For starters, we need to listen to the automatic terminal information service (ATIS) and get the latest weather briefing. Then we can call the tower and request clearance for departure.

"John Wayne Tower, this is Cessna 22287 with information Charlie, requesting departure."

"Cessna 287, you are cleared for immediate departure, runway 19 right."

"Roger, 22287 departing 19 right."

Okay, we've got clearance. Go ahead and give it full throttle. The aircraft will want to shift a little to the left from the torque of the engine. When it does, just add a little right rudder, using the + (plus) or Enter key on the numeric keypad, to compensate. (Remember that pressing keypad 5 will center the rudder.) Rotate at 80 knots, and then raise the gear. Try to keep the climb speed at 80 knots. Remember, because we are now using more realistic settings, you are going to need continual back pressure to keep the nose in a climb—or use the elevator trim (press keypad 1 and keypad 7) to adjust the climb attitude. Let's keep this heading until we reach our target altitude of 4500 feet or until ATC tells us differently.

Get the Weather Briefing and Take Off

1. Press C to listen to the latest ATIS weather briefing.
2. Press . (period) to release parking brakes if they are set.
3. Hold down keypad 9 to apply full throttle.
4. Press the + (plus) or Enter key on the numeric keypad to apply a little right rudder to keep the aircraft centered down the runway. (Remember to use keypad 5 to bring the rudder back to a neutral position.)

5. Rotate and lift off at 80 knots.
6. After lifting off the runway, press G to raise the gear.
7. Use keypad 7 and keypad 1 to trim the climb attitude of the aircraft for an airspeed of 80 knots.

As we are climbing, we get a call from John Wayne Tower:

"Cessna 287, continue climb to 4500 feet. Contact departure control."

"287, roger..."

"...Departure control, this is Cessna 22287, climbing to 4500."

"Cessna 22287, continue climb to 4500 and turn to a heading of 270 degrees."

"Roger, 287, 270 degrees for 4500 feet."

Chapter 4: Instrument Flight

Aeronautically Speaking:
Actually, all IFR flights initiate with a filed flight plan. The flight plan gives the tower information about the aircraft and the planned departure time. During the entire flight, ATC stays in touch with the pilot. As ATC watches the aircraft signal on the radar screens, it also sees an attached block of information that displays the type of aircraft, the identification number, the airspeed, and the altitude. This information allows ATC to coordinate the movement of the aircraft with other aircraft that are in the area, which helps prevent midair collisions. Thus, IFR flight requires not only relying on instruments but also maintaining constant contact with ATC.

Because ATC has a copy of our flight plan and knows where we are going, it will give us directions for the best route during our flight. It will try to stick to the flight plan we have filed but might make some changes to keep us away from other aircraft or from some nasty weather. But before we make any maneuvers, we should recalibrate the directional gyro, because it has probably drifted by now. Then let's go ahead and make a climbing turn to a heading of 270°. Keep the airspeed at 80 knots during the turn, and, after reaching the desired heading, continue to the target altitude of 4500 feet. Remember to rely on the directional gyro, the artificial horizon, and the turn and bank indicator for heading and attitude information.

As we pass through 4000 feet, we should be "on top," with the clear blue sky above and the puffy white clouds below. When we reach 4500 feet, level off, reduce the throttle to 80-percent power, and trim the aircraft for straight and level flight. As soon as the aircraft is level at 4500 feet, we should contact departure control and then pause the simulator while you learn to get your bearings by using the NAV radios and the OBI instruments.

Make a Climbing Turn to 270°

1. Press D to recalibrate the directional gyro.
2. Make a standard-rate right turn to a heading of 270° at an altitude of 1000 feet during your climb.
3. Maintain an airspeed of 80 knots during the climb to an altitude of 4500 feet.
4. At 4500 feet, level off and reduce the throttle to 80 percent, and use keypad 7 and keypad 1 to trim the aircraft for straight and level flight.
5. Press P to pause the simulator.

"Departure control, Cessna 22287 is at 4500 feet."

"Roger, 287, contact Los Angeles Center, and good day..."

"...Los Angeles Center, this is Cessna 22287 with you at 4500 feet, heading 270 degrees."

"287, LA Center, continue on present course."

"287, roger."

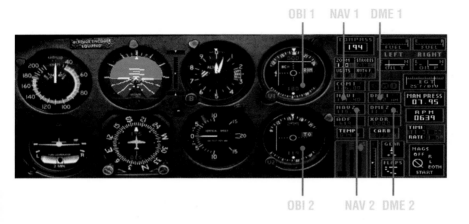

The Cessna navigation instruments.

Navigation Instruments

The instruments that we will be using for navigation are the omni-bearing indicator (OBI), the navigation radio (NAV), and the distance measuring equipment (DME). We will also rely heavily on a piece of equipment that is located on the ground—the very high frequency omnidirectional range (VOR).

The Learjet navigation instruments.

Chapter 4: Instrument Flight

The VOR is a ground-based transmitter that sends a directional signal along magnetic headings, or radials. The NAV radio receives this signal and communicates to the OBI which magnetic radial the aircraft is on. It is then up to the pilot to adjust the OBI degrees until the vertical needle is centered. Doing this will tell the pilot which magnetic radial the aircraft is on relative to the position of the VOR. The DME provides additional information by measuring the aircraft's distance from the VOR. It also uses the changes in the distance information during flight to measure the ground speed of the aircraft.

How the OBI reacts to a VOR signal.

The OBI 1 shows that the aircraft is currently on the 187° radial of the VOR. The "To" flag shows that the aircraft would have to fly at a heading of 187° in order to fly directly to the VOR.

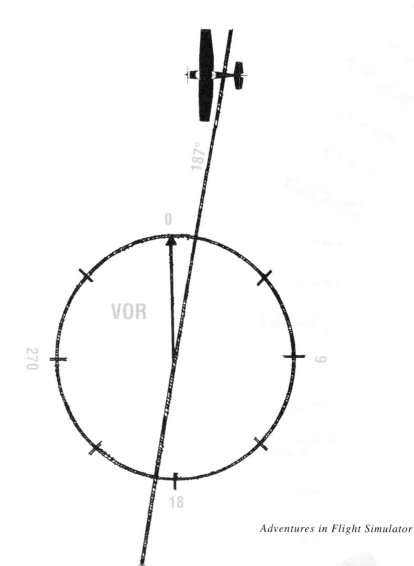

As you see in this example, the aircraft is flying a heading of 270° but is on the 187° radial of the VOR. The pilot had to tune the OBI until the needle was centered on the middle circle. The "To" in the To-From indicator shows that the heading displayed at the top of the OBI is the heading "To" the VOR.

Look at the Los Angeles sectional and find the Seal Beach VOR. Our NAV 2 radio and OBI 2 are tuned to this VOR. Now look at the Santa Catalina VOR. The NAV 1 radio and OBI 1 are tuned to this VOR. When you have finished looking at the map, let's return to flying.

Review the Los Angeles Sectional

1. Look at the Los Angeles sectional chart, and notice the location of the Seal Beach VOR and the Santa Catalina VOR.
2. Press P to unpause the simulator, and keep the map in your lap for now.

We need to find the exact location of our aircraft by using the Seal Beach and Santa Catalina VORs. The NAV 2 radio is tuned to Seal Beach, so adjust OBI 2 until "From" is indicated in the To-From indicator and the needle is centered in the gauge. Notice which way the needle is moving as we continue flying. If the needle is moving to the left, adjust OBI 2 until the needle is halfway between the right side of the instrument and the center. (If the needle is moving to the right, it means that we are flying in the wrong direction!) Now, while the needle of OBI 2 is moving closer to the center, quickly adjust OBI 1 until the needle is in the center of the instrument. With good timing, you should be able to get both needles centered almost simultaneously. Now pause the simulator and we'll calculate our exact location.

Find Headings from Each VOR

1. Press the V key and then the 2 key at the top of the keyboard to activate the OBI 2 instrument. Use the + (plus) and – (minus) keys at the top of the keyboard to adjust the heading until the vertical needle is between the center and the right side of the instrument and "From" is in the To-From indicator.

2. Press the V key and then the 1 key at the top of the keyboard to activate the OBI 1 instrument. Use the + (plus) and – (minus) keys at the top of the keyboard to adjust the heading until the vertical needle is centered and "From" is in the To-From indicator.

3. Press P to pause the simulator when both OBIs are centered as well as possible.

Where Are We?

Look at the heading for OBI 2. Using a pencil and a ruler, draw a line from the center of the Seal Beach VOR to the radial for the heading on OBI 2. The marks on the outside ring of the VOR are in 5-degree increments. For example, if the heading at the top of OBI 2 is 190°, look for the radial marked "21" (for 210°) and count the marks, subtracting 5° for each mark, until you reach the mark for 190°. When you find the appropriate radial mark, extend the pencil line past the circle of the VOR another 6 to 8 inches on the map. Do the same for the OBI 1 heading with the Santa Catalina VOR. (The two pencil lines on the existing Los Angeles sectional from the Santa Catalina and Seal Beach VORs serve as examples of how your lines should look.) The point at which these two lines meet (the intersection angle) represents the location of your aircraft. Now, that wasn't too bad, was it? Having a pause key is a great help. Next time, though, we'll do the instrument navigation without pausing the simulator.

Wing Tip:
You can also use the mouse as a quick way to adjust your OBI headings. Clicking on the rightmost number of the heading increases your heading by 1°. Clicking on the leftmost number of the heading decreases your heading by 1°. You do not have to press V and the OBI number to activate the OBI when you use the mouse to adjust your headings.

Pinpoint the Aircraft's Position

1. Draw a line from the center of the Seal Beach VOR, through the radial mark for the OBI 2 heading, about 8 inches past the VOR circle on the sectional chart.
2. Draw a line from the center of the Santa Catalina VOR, through the radial mark for the OBI 1 heading, about 8 inches past the VOR circle on the sectional chart.
3. Note the intersection angle, which is the position of the aircraft.

Take another look at the sectional chart. You'll notice a line extending from the Seal Beach VOR to the Santa Catalina VOR. The middle of this line is labeled "V21," showing that it is a designated route of travel called "vector 21." To the right of this designation you'll see "←202°", showing that the route in the direction of the arrow is at a heading of 202° from the Seal Beach VOR. For our approach to Santa Catalina we will need to follow this vector.

Let's unpause the simulator and set OBI 2 for our approach to Santa Catalina airport. Press V and then 2 at the top of the keyboard to activate OBI 2. Use the + (plus) and – (minus) keys at the top of the keyboard to tune OBI 2 until the top heading reads 202°. We should recalibrate the directional gyro again. Continue in straight and level flight until the needle on OBI 2 begins to move. We'll get in touch with LA Center and make a request for vector 21 to Santa Catalina. When

Chapter 4: Instrument Flight

the needle reaches halfway between the center and the right side of the instrument, make a standard-rate left turn to a heading of 202°. By the time we come out of the turn, the needle of OBI 2 should be centered. Continue with straight and level flight. If the needle is not completely centered, adjust the heading to intercept the signal. If the needle is to the right of center, we need to turn a little to the right until the needle is centered. When it is centered, readjust the heading to 202°. Once we are on course, note our distance from Santa Catalina by looking at the DME.

"LA Center, Cessna 22287 requesting vector 21 to Santa Catalina."

"287, LA Center, intercept vector 21 and proceed to Santa Catalina. Make request for approach altitude when ready."

"22287, intercepting vector 21, thank you."

Intercept Vector 21

1. Press D to recalibrate the directional gyro.
2. Press V and the 2 key at the top of the keyboard and use the + (plus) or − (minus) keys at the top of the keyboard to tune OBI 2 to a heading of 202°.
3. Continue straight and level until the OBI 2 needle is halfway between center and the right side of the instrument, and then make a standard-rate left turn to a heading of 202°.
4. Make any necessary heading adjustments to center the OBI 2 needle.
5. Note the distance reading on the DME.

Our distance from Santa Catalina should be somewhere around 20 miles. When we reach a distance of 15 miles, we'll contact ATC and begin our descent into the Santa Catalina airport. Until we reach the 15-mile mark, feel free look at the aircraft in Spot view as it flies above the cloud layer. But don't spend too much time looking around. At 140 knots, it only takes a few minutes to travel 5 miles. This would also be a good time to save the flight as a situation. Then you can always come back and practice the approach into the Santa Catalina airport.

Take Time for Viewing

1. Press S twice to switch to Spot view.
2. Press Shift and the keypad keys to look around as desired.
3. Press S to return to Cockpit view.
4. Open the Options menu, select Save Situation, enter the description **Before approach into Santa Catalina**, and save the flight as a situation named **CH4FLT1A**.

Approach Into Catalina

"LA Center, Cessna 22287 at 15 miles from Santa Catalina on vector 21 requesting approach into Santa Catalina."

"287, maintain heading and descend to 2000 feet. Contact Catalina Tower."

"287, descending to 2000, good day....Catalina Tower, this is Cessna 22287, inbound at 15 miles on vector 21, descending to 2000. Requesting approach to runway 22."

"22287, Catalina Tower, cloud ceiling is at 2500 feet. Continue descent to 2000 feet and report when you have a visual."

"22287, roger."

When the DME on the NAV 1 DME indicator shows us at 15 miles from the airport, apply the carburetor heat, reduce the throttle to 30-percent power, and begin a descent into the clouds, which we'll enter at around 4000 feet. Continue the descent through the clouds. Rely on the instruments for staying on a constant heading of 202°, and make sure that the descent does not exceed the maximum airspeed for the aircraft. Because our vector is pointed at the Santa Catalina VOR and the airport is to the right of the VOR, we will need to make a right turn as soon as we come out of the clouds. We should start encountering patchy clouds at 2500 feet, through which we will be able to visually establish where the island is. When we reach 2000 feet, level off and increase the throttle to 50-percent power.

Descend Through the Clouds

1. Press H to apply carburetor heat when the DME on the NAV 1 DME indicator reads 15 miles.
2. Use keypad 3 to reduce the throttle to 30-percent power, and begin your descent through the clouds.
3. Use keypad 9 to increase the throttle to 50 percent, and level off at 2000 feet.

"Catalina Tower, Cessna 22287 at 2000 feet with visual, on straight-in approach for runway 22."

"22287, continue with approach, you are cleared to land, runway 22."

"287, roger, cleared for landing."

We should be able to see the runway off in the distance. Stay on a course of 202°, but watch as the runway gets closer. When the runway appears just off to your right, turn to a heading of 220° and then adjust your heading to make a straight-in approach to the runway. As soon as we are lined up, drop the flaps to 30°, reduce the power to 30 percent, and lower the gear. Remember, the runway is at an altitude of 1602 feet, so we don't have much distance to descend. As soon as we pass the end of the runway, cut the power, and then gently flare for the landing.

Land at Catalina Airport

1. Remain at 2000 feet and look for the runway.
2. When the runway appears just off to your right, turn right to a heading of 220° and adjust your heading for a straight-in approach to the runway.
3. Use keypad 3 to reduce power to 30 percent, and press F7 to lower the flaps to 30°.
4. Press G to lower the gear.
5. Use keypad 3 to cut the power when you are over the threshold, and flare for landing.

"Catalina Tower, Cessna 22287 is requesting taxi to the restaurant, and would you please close our flight plan?"

"22287, we'll take care of the flight plan. Cleared to taxi for lunch. Enjoy your meal."

"287, roger. Thanks."

Flight Summary

If you completed this adventure successfully, you're ready for some more IFR flying. Because we covered quite a bit of new information during this flight, feel free to fly through the situation a few times to become comfortable with using the instruments for navigation. If you are feeling really confident, try the flight again without pressing pause to make instrument changes or to look at the map. Do it all while the plane is flying.

Situational Awareness

Before we go on our next IFR flight, let's talk a little about situational awareness, or the ability to know where and in what position you are at any single point in time. Situational awareness can be difficult under IFR because it requires full reliance on the instruments and the pilot's ability to interpret what the instruments are saying. But you can ask yourself several questions during a flight to help stay situationally aware.

- **What do the instruments say?** This is the most important question you can ask yourself during an instrument flight. In the real world, many pilots have met their doom because they thought they knew more than the instruments. It's very easy to second-guess an instrument and assume that it is failing. Actually, Flight Simulator pilots have an easier time adapting to using the instruments because the software is their sole source of information on how the aircraft is performing and what situation it is in. Real-world pilots also rely on their senses and want to believe what they are "feeling" over what the instruments are saying. A good pilot knows that the instruments have priority.

- **What is the current ground separation?** Most flying accidents and crashes occur because the pilot doesn't realize how close the ground is. In IFR flights, it is very important to scan the altimeter regularly and often. During an approach to an airport it is easy to fixate on the glide slope (ILS approach indicators on the OBI 1 instrument) and forget about the other instruments. Regardless of the situation, always remember to keep an eye on the altimeter.

- **What is the environment?** Are there buildings nearby? Mountains? Other aircraft? It is important to ask these questions when flying IFR. Because the view perception in Flight Simulator is limited (compared to the real world) it can sometimes be difficult to anticipate how quickly you will arrive at a building or a mountain while making a turn. Even when you are in the clouds, it is a good idea to keep a mental picture of where the obstacles are. This is why using the instruments precisely is so important for IFR flying. A few degrees of error or a few hundred feet of altitude can mean the difference between an enjoyable flight and one that meets an abrupt end.

Chapter 4: Instrument Flight

Reno to Oakland

Because you made such great progress with the instrument flight to Santa Catalina island, let's get a taste of the life of a commercial pilot. During this flight from Reno, Nevada, to Oakland, California, we will use the Autopilot. As a result, we will experience the boredom of an automated instrument flight and the sheer terror and excitement of an instrument approach into an airport under poor weather conditions.

Flight Plan: We will fly the Learjet out of Reno Cannon International airport to Oakland International airport. The weather in Reno will be clear for our 5:15 p.m. departure, but by the time we get to Oakland it will be getting dark. And there is a good chance that we will encounter some marine fog over Oakland. We will take advantage of the Autopilot features of Flight Simulator and will spend some more time coordinating our position by using the OBIs. The flight will end with a full instrument ILS approach into Oakland at night.

Flight Settings

Aircraft	Learjet 35A
North/South Latitude	N039°29'51.3642
East/West Longitude	W119°45'59.7603
Altitude	4415 feet
Heading	344°
Season	Autumn
Time of Day	5:15 p.m.

(continues on next page)

Flight Settings *continued*
Weather—San Francisco Area
Beginning Latitude	N038°00'00.0000
Beginning Longitude	W123°00'00.0000
Ending Latitude	N034°00'00.0000
Ending Longitude	W123°00'00.0000
Width	50 miles
Course	000°
Speed	0 knots

Clouds
Base	800 feet
Tops	2000 feet
Coverage	Overcast
Deviation	300 feet

Temperature
Altitude	10,000 feet
Daytime Temperature	25°F

Auto Coordination	Off

Realism and Reliability
Flight Control Realism	1
Elevator Trim	On
Gyro Drift	On
Airframe Damage from Stress	On
Engine Stops When Out of Fuel	On
Instrument Lights	On
Lights Burn Out	On
Fuel Tank Selector	On

Navigation Instruments
NAV Radio 1	115.5
OBI 1	200°
NAV Radio 2	116.0
COM Radio 1	128.5

Enter the Flight Settings

1. Open the Options menu.
   Select the Learjet 35A as your aircraft.

2. Open the World menu.
   Select Set Exact Location.
   Set North/South Lat. to **N039*29'51.3642** and East/West Lon. to **W119*45'59.7603**.
   Set Altitude to **4415** and Heading to **344**.

3. Open the Sim menu, and be sure Auto Coordination is not selected.
   Select Realism and Reliability.
   Leave Flight Control Realism at 1, and select Elevator Trim, Gyro Drift, Airframe Damage from Stress, Engine Stops When Out of Fuel, Instrument Lights, Lights Burn Out, and Fuel Tank Selector.

4. Open the World menu.
   Select Set Time and Season.
   Set Season to Autumn.
   Set Exact Time to 17:15.

5. Open the World menu again, and select Weather.
   Select Add Area.
   Type **San Francisco** in the Area Name box and press Enter.
   Set Beginning Lat. to **N038**, Beginning Lon. to **W123**, Ending Lat. to **N034**, Ending Lon. to **W123**, and Width to **50**. Leave Transition, Course, and Speed at 0.

6. Select San Francisco from the Weather Area menu.
   Select Create in the Cloud Layer box.
   Set clouds to have a Base of **800**, Tops of **2000**, a Coverage of Overcast, and a Deviation of **300**.

7. Select Temperature and create a temperature layer with a Temperature Altitude of **10000** and a Daytime Temperature of +025.

8. Open the Nav/Com menu.
   Select Navigation Radios.
   Set the NAV 1 frequency to **115.5**, OBI 1 to **200**, and NAV 2 frequency to **116.0**.

9. Open the Nav/Com menu again.
   Select Communications Radio.
   Set the COM 1 frequency to **128.5**.

10. Open the Options menu.
    Select Save Situation.
    Save this situation as **CH4FLT2**.

Wing Tip:
Remember that the Learjet is very sensitive, and it handles a lot differently than the Cessna. One of the main characteristics of the Learjet is that it goes very fast and is therefore difficult to slow down. Remember this when we make our final approach into the Oakland airport.

Because we will be using the San Francisco sectional from the *Flight Simulator Pilot's Handbook* and may need to make some pencil marks, it would be a good idea to make a few photocopies of the map. Clip them onto your kneeboard so you can use them during the flight.

Pretakeoff Considerations

Before we take off, we need to look over a few things in preparation for our flight. To begin, we need to look at the San Francisco sectional in the *Flight Simulator Pilot's Handbook*. Notice where Reno Cannon airport is (our starting point), and then look for the location of the Oakland International airport (our end point). We will be using various VORs for navigation during our trip. The first VOR we use will be the Hangtown VOR. The next will be the Manteca VOR. From there, we will make an instrument landing system (ILS) approach using the Oakland ILS runway 29 approach plate.

To make the flight easier to follow, you can use the chart and draw a line from Reno to the Hangtown VOR and then from the Hangtown VOR to the Manteca VOR. Draw a line from the center of the Manteca VOR along the 229° radial (one major notch down from the radial marked 24) and out toward the Moffett NAS airfield. Next, draw a line from the leftmost runway at Oakland, at the same heading as the runway, toward the line that you drew from the Manteca VOR. At the point where the lines intersect, we will make a turn to intercept the ILS at Oakland.

In preparation for our flight, we have already set our NAV 1 radio to the Hangtown VOR, the OBI 1 to a heading of 200°, the NAV 2 radio to the Manteca VOR, and the COM 1 radio to the ATIS frequency in Oakland. This will help to cut down on our need to adjust the radios during the flight.

Take Off

1. Press . (period) to disengage the parking brakes if they are on.
2. Press D to recalibrate the directional gyro.
3. Hold down keypad 9 to apply full throttle.
4. Rotate at 130 knots, and then press G to raise the gear.
5. Make a steep left turn during the climb to a heading of 200°.
6. Use keypad 1 and keypad 7 to trim the aircraft, if necessary.
7. At 12,000 feet, use keypad 3 to reduce the throttle to a fan level of 80 percent, and trim the aircraft for level flight.

"Nevada Center, this is Learjet Foxtrot Sierra Five Lima, with you at twelve thousand at two hundred degrees."

"Lear Five Lima, proceed to Hangtown as filed. Climb and maintain flight level two six."

"Five Lima, roger, Hangtown, flight level two six."

Before starting our climb to 26,000 feet, adjust the heading on OBI 1 until the needle is centered. The heading displayed at the top of the OBI should be our heading for flying to the Hangtown VOR. Recalibrate the directional gyro again, and then turn to the heading indicated on the OBI. We should be around 54 miles from the VOR, as indicated by the DME on the OBI. As soon as we are on the correct heading, apply full throttle again and climb to 26,000 feet.

Adjust the OBI

1. Press V and 1 and then use the + (plus) and – (minus) keys to adjust the OBI 1 heading until the needle is centered.
2. Press D to recalibrate the directional gyro.
3. Turn to the heading indicated on OBI 1.
4. Use keypad 9 to apply full throttle, and begin the climb to 26,000 feet.

Wing Tip:
The heading hold feature of the Autopilot will lock itself onto the heading of the magnetic compass. Because the directional gyro has a tendency to drift, this is not a reliable device for holding a constant heading.

Autopilot: The Pilot's Assistant

"Lear Five Lima, contact San Francisco Center. Good day."

"Roger, Nevada Center, this is Learjet Five Lima. Thank you…"

"…San Franciso Center, this is Learjet Foxtrot Sierra Five Lima, climbing for flight level two six to Hangtown."

"Sierra Five Lima, San Francisco Center has you on radar. Continue climb to flight level two six."

"Roger, Five Lima."

Start leveling off when we pass through 25,500 feet by lowering the nose and reducing the throttle to a fan setting of 80 percent. (The turbine speed gauges will show 90 percent.) We should be at 26,000 by the time you get the Lear leveled off. Recalibrate the directional gyro again. The DME should show our distance at around 36 miles from the Hangtown VOR. Now it is time to rely on our automated assistant, the Autopilot system. From the Nav/Com menu, select Autopilot. Enter the current heading in the Heading box, and enter **26000** in the Altitude box. Then change "Disconnected" to "Connected" to turn on the Autopilot. Click OK, and the aircraft will be under the control of the Autopilot system.

Set the Autopilot

1. At 25,500 feet, begin leveling off, and reduce the throttle to a fan setting of 80 percent.
2. Press D to recalibrate the directional gyro.
3. Select the Nav/Com menu, and then select Autopilot. Enter **26000** in the ALT (altitude) Hold box, and enter the current heading in the HDG (heading) Hold box.
4. Select Connected (On) from the Autopilot Switch box.

Our distance to the Hangtown VOR should be around 28 miles. While the Autopilot has control of the aircraft, we can utilize the time to look at the engine instruments and check our fuel level. It would also be a good time to look at the various internal and external views. Feel free to take a flight photograph if you want. Then take a look at the indicated distance to the Hangtown VOR on the DME. When we reach a distance of 10 to 15 miles from the VOR, go ahead and switch over to the OBI 2 display.

Check the Engine Instruments

1. Press Tab to display the engine instruments, and check the fuel level.
2. Press Tab again to display the radio instruments.
3. If desired, check out the various views or take a flight photograph.
4. Press Shift-Tab to select the OBI 2 display when the DME reads 10 miles.

As part of our preflight, we entered the frequency for the Manteca VOR for NAV 2 and OBI 2. The OBI 2 display is already "linked" to the Manteca VOR, and the DME will show that we are around 68 to 73 miles from the VOR. Recalibrate the directional gyro again, and then adjust OBI 2 until the needle is centered. The heading at the top of the OBI instrument should be our new heading for flying to the Manteca VOR. Turn off the Autopilot, and then bank left to the new heading. As you make the turn, you'll notice that the needle will again move slightly off to one side of the OBI. After leveling off at the new heading, readjust the OBI again until the needle is centered. Again, make a slight turn to the new heading. We should need to adjust our heading by only a few degrees. When we are level on the new heading, go ahead and set the Autopilot, just as we did earlier, and turn it back on.

Change the Heading

1. Press D to recalibrate the directional gyro.
2. Press V and 2, and then use the + (plus) and – (minus) keys to adjust OBI 2 until the needle is centered with "To" displayed in the "From-To" indicator.
3. Press Z to turn off the Autopilot.
4. Turn to the new heading indicated on OBI 2.
5. Adjust the OBI 2 heading again until the needle is centered.
6. Make a slight turn to the new heading indicated on OBI 2.
7. Enter the new heading in the Autopilot dialog box, and turn on the Autopilot just as you did earlier.

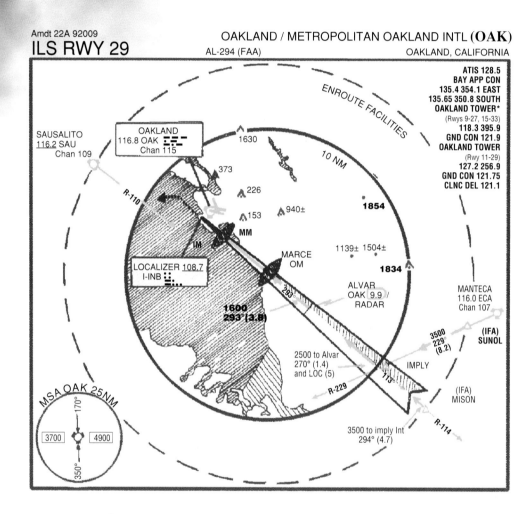

Midflight Approach Planning

"San Francisco Center, this is Lear Five Lima, enroute to Manteca VOR, at flight level two six."

"Roger, Five Lima. Proceed as planned."

"Five Lima."

Because we are over 55 miles away from the Manteca VOR, we have a little time to plan our approach into the Oakland airport. However, you might want to press P to pause the simulator now to give yourself still more time to become comfortable with the planned approach. Take a look at the approach plate for Oakland ILS runway 29.

On the right side of the chart, you will notice the word *Manteca*, with *116.0 ECA* below it. (ECA is the three-letter identifier for the Manteca VOR.) The triangle shape below and to the right of the words represents the Manteca VOR. Follow the line from the VOR, and you will see 3500 over 229°. This tells us that we need to be at a heading of 229° and at an altitude of 3500 feet when we fly toward the ILS approach for Oakland runway 29. The line then turns right and follows the middle of the giant arrow shape. When our Learjet is within the giant arrow (the range of the ILS radio signal), we will be able to receive the ILS signal, which will allow us to stay lined up with the runway. In the middle of the line you will notice that the heading for the approach is 293°. Just to the left of the heading, you can read that the aircraft should be at 1600 feet at a heading of 293°.

Just past the 293° heading is the lens-shaped outer marker beacon. The OBI will beep at us and will display a blue *O* when we fly over the top of the outer marker. Just in front of the runway is the middle marker, and then the inner marker. In the box next to the middle marker, you can find the frequency for the ILS approach, which is 108.7. We can do some preplanning for our approach by setting OBI 1 and the NAV 1 radio for the ILS. Go ahead and unpause the simulator and then set the NAV 1 radio frequency to 108.7. We don't need to set the OBI heading because the heading needle and glide slope needles will automatically line up for the approach.

Examine the Approach Plate

1. Press P to pause the simulator, and examine the approach plate.
2. Press P again to unpause the simulator.
3. Use the Nav/Com menu to change the NAV 1 radio frequency to 108.7.

As we get closer to the Manteca VOR, take another look at the engine instruments and check the fuel level again. Switch back to the radio instruments. When the DME shows that we are 10 miles away from the Manteca VOR, remember that we will be changing our heading to 229°, as shown on the approach plate. This will allow us to intercept the ILS approach. This would be a good time to save the situation. Because we are going to get a little busy with the approach and landing, it would be nice to be able to come back and practice the approach from this point without having to fly the entire flight again. (And if you crash into the runway, you can always have another chance!)

Save the Situation

1. Press Tab to examine the fuel level.
2. Press Tab to display the radio instruments again.
3. Use the Options menu to save the situation as **CH4FLT2A**.

"Lear Sierra Five Lima, after passing Manteca VOR, make a right turn heading two two nine degrees, and then descend to flight level one zero."

"Roger, San Francisco Center. Sierra Five Lima turning to two two nine, descending to flight level one zero."

As we draw closer to the Manteca VOR, the OBI needle will start to swing off to one side. When we pass over the top of the VOR, the "To" text will change to "Off" and then to "From" because our OBI heading now indicates the heading away from the VOR instead of to the VOR. Recalibrate the directional gyro and then turn off the Autopilot. Make a right turn to a new heading of 229°. After leveling out at the new heading, reduce the throttle to a fan setting of 50 percent. Let the nose of the aircraft drop below the horizon for a standard descent, but use the trim to keep from descending too steeply. While descending at this heading, change the frequency of the NAV 2 radio to the frequency of the Oakland VOR (displayed on the approach chart in the box at the top left of the circle—116.8 is the frequency). Now, change the OBI 2 heading display to a heading of 293°. Yes, this is the same heading as the ILS approach. By setting the OBI 2 to the same heading, we will be able to tell when we are getting close to the approach path of the runway.

Prepare for Approach

1. Note that "To" changes to "From" on the OBI 2 instrument, which indicates that you have flown over the Manteca VOR.
2. Press D to recalibrate the directional gyro.
3. Press Z to turn off the Autopilot, and turn to a new heading of 229°.
4. Use keypad 3 to reduce the throttle to 50 percent, after leveling off.
5. Use the Nav/Com menu to change the NAV 2 radio frequency to 116.8.
6. Press V and 2, and then use the + (plus) and – (minus) keys to adjust the OBI 2 heading to 293°.

Our DME should be showing a distance of 30 miles (or more) from the Oakland VOR. Continue the descent at the heading of 229°. The weather report for Oakland mentions that there is a low fog layer over the Bay Area with a ceiling of 800 feet, but that's nothing to be concerned about because we are relying on instruments.

To aid our descent and slow down the Learjet for the approach, raise the spoilers. The spoilers will create a lot of drag on the aircraft and will restrict the airflow over the wings. Keep the spoilers out until the airspeed reaches 200 knots. Apply 8° of flaps, and remember to compensate for the additional lift by keeping the nose down.

For our approach we are going to use a special visual tool to guide us through the fog over Oakland. We will turn on the electronic flight instrument systems (EFIS) and command flight path display (CFPD), which will calculate and project a flight

path on the front windscreen for us to follow on our approach to the runway. From the Nav/Com menu, select EFIS. Select Master Switch, Lock to ILS for Landing Approach, and Plot Intercepting Path. Set the Type to Rectangles, the Density to Thin, and the Range to Medium. Then click OK. You will be able to see the red rectangles for the flight path. Because we told the system to plot the intercepting path, we can see the correct path for meeting the approach path. If you are off to one side or if the altitude is off, that's okay. We will continue to rely on the instruments for the intercept path.

Set up the EFIS

1. Press the / (slash) key when the DME shows 30 miles to extend the spoilers (or click on EXT in the SPL box in the radio panel).
2. Press the / (slash) key again or click on RET to retract the spoilers when the airspeed reaches 200 knots.
3. Press F6 to apply 8° of flaps.
4. Open the Nav/Com menu, and then select EFIS/CFPD Display.
5. Select EFIS Master Switch, Lock to ILS for Landing Approach, and Plot Intercepting Path. Then set the Type to Rectangles, the Density to Thin, and the Range to Medium.

On Final Approach to Oakland

"Lear Sierra Five Lima, continue your descent to 3500 feet. Intercept the localizer at one zero eight dot seven, and contact approach control."

"Roger, San Francisco Center. Will intercept localizer at one zero eight dot seven. Switching to approach. Thanks..."

"...Approach control, this is Learjet Foxtrot Sierra Five Lima, approaching localizer for runway two niner."

"Sierra Five Lima, approach control has radar contact. Continue to descend to the localizer. You are cleared for landing. Please be advised that a marine layer is over the airfield with a ceiling of eight zero zero feet."

"Roger, approach. Weather acknowledged. Five Lima cleared for landing, runway two niner."

Reduce the power to a fan setting of 40 percent and continue on the heading of 229°. Recalibrate the directional gyro. As we get closer to the approach path, you will notice that the OBI needle will start to move. When this happens, go ahead

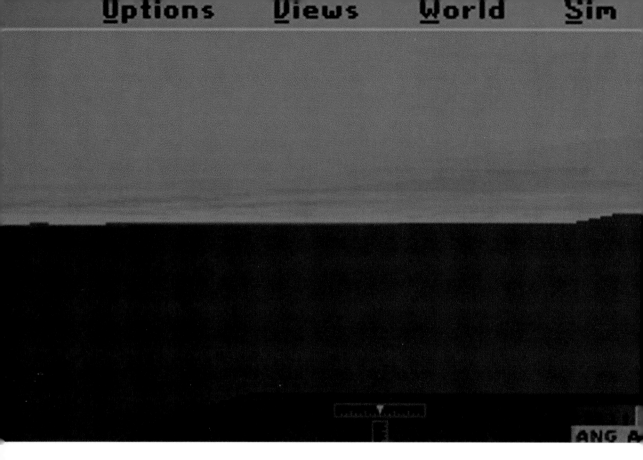

and start a right turn to a new heading of 293°. Try to coordinate the turn so that the aircraft will finish the turn lined up in the center of the red rectangles. (You might have to make a steep right turn.) If you are slightly off to one side, use minor adjustments to gently bring the aircraft into the rectangles. Switch to the OBI 1 display, and you will be able to see the glide slope for the ILS approach. Notice how the positions of the needles relate to the positions of the red rectangles. If the needles are off to the left and below the center of the OBI, the aircraft is too far to the right and too high.

Make the Final Approach

1. Use keypad 3 to reduce the throttle to a fan setting of 40 percent.
2. Make a steep right turn to a heading of 293° when the OBI needle starts to move.
3. Try to end the turn in the middle of the red rectangles.
4. Adjust as necessary for flying down the center of the red rectangles.
5. Press Shift-Tab twice to switch to the OBI 1 display.

Lower the flaps to 20°, and then drop the gear. Compensate for the change in attitude, and try to keep the aircraft in the red rectangles. You might briefly be able to see the airport up ahead, but once we pass through 2500 feet, we will start to enter into the fog layer. If you are having difficulty staying at the correct altitude, make minor adjustments to the throttle. Increase the throttle for more lift, and decrease it for less lift. At 7 miles out (according to the DME), we will fly over the outer marker beacon. The airspeed should be around 165 knots. Remember to use a very light touch on the controls as you keep the aircraft in the red rectangles. As we come out of the fog, the runway should be directly in front of us. As soon as we pass over the end of the runway, reduce the throttle to a fan setting of 20 percent and let the aircraft settle gently down to the runway. Right after the wheels touch down, reduce the throttle to full reverse and keep it there until the aircraft slows down to below 40 knots. You can then use the brakes to stop.

Chapter 4: Instrument Flight

Land the Aircraft

1. Press F7 to lower the flaps to 20°, and press G to lower the landing gear.
2. Use the throttle to control the descent rate, and stay in the red rectangles.
3. Use keypad 3 to reduce the throttle to a fan setting of 20 percent when you are over the end of the runway.
4. Gently set down on the runway.
5. Hold down keypad 3 to apply full reverse throttle until the airspeed is 40 knots.
6. Press . (period) to apply the brakes for stopping.

Whew! If you made it through successfully the very first time, pat yourself on the back. However, if it took you a few tries to get it right, don't feel bad. This is a very difficult flight if you're not familiar with the instruments or the Learjet. But when you are able to complete the flight, you can truly call yourself a professional Microsoft Flight Simulator pilot!

In this chapter, we covered the following topics:

Using realism settings
Flying with separate rudder and aileron controls
Using VORs
Using the NAV radios and OBIs for navigation
DME information
Using a sectional map
Pinpointing position with two VORs
Situational awareness
Autopilot controls
Landing with the EFIS/CFPD system
Glide slope ILS approaches

Chapter 5

Graphics, Scenery, Pictures, and Video

"Wow! Where'd you get the pictures?" the pilot's friend exclaims.

"Oh, I took those flying over Seattle. Here's a shot of the Learjet I was flying."

"You flew a Learjet?"

"Sure did! And I have some great videos of the flight, too. Come on, I'll play them for you."

The Art of Using Art

You can learn a lot about flying, aerodynamics, instrumentation, and navigation by using Flight Simulator. But the real joy of flying Flight Simulator is provided by the program's realistic graphics, which enhance the experience tremendously.

To help demonstrate some of the major advancements that have been made in Flight Simulator 5, the discussion in this chapter divides the new features into four major categories: graphics, scenery, pictures, and video. And what better way to look at the effect of the various graphics settings than during an actual flight?

Flight Plan: For this flight, we'll use the Learjet and will depart from runway 26 at Olympia Airport in the state of Washington. The flight will be mostly straight and level, with some turns, and will demonstrate the effect of the various graphics modes and settings on the appearance of the graphics and on the performance of Flight Simulator.

✛ ✛ ✛

Flight Settings

Location	Olympia Airport, Runway 26
Aircraft	Learjet 35A
Auto Coordination	On
Season	Winter
Time of Day	4:00 p.m.
Weather	Clouds
Base	10,000
Tops	12,000
Coverage	4/8, Scattered

Enter the Flight Settings

1. Open the Options menu.
   Select Aircraft.
   Select Learjet 35A.

2. Open the World menu.
   Select Airports.
   Select USA-Seattle from the Choose A Scenery Area menu.
   Select Olympia-Runway 26 from the Choose The Airport You Want To Fly From box.

3. Open the World menu again.
   Select Set Time and Season.
   Set Season to Winter.
   Select Set Exact Time.
   Set Hours to **16**, and set Minutes to **0**.

4. Open the World menu again.
   Select Weather.
   Select Edit in the Cloud Layers box.
   Set Base to **10000**.
   Set Tops to **12000**.
   Set Coverage to Scatter 4/8.

5. Open the Sim menu and be sure Auto Coordination is selected.

6. Open the Options menu.
   Select Save Situation, and save this situation as **CH5FLT1**.

Chapter 5: Graphics, Scenery, Pictures, and Video

Wing Tip:
There are three important rules to follow when flying the Learjet:
* *Fly with a light touch on the controls, and use slow, steady movements to adjust them.*
* *Do not exceed the maximum operating speed (460 knots), which is easy to do, even in straight and level flight. When the "Overspeed" message appears, reduce the throttle and slow down. Otherwise, supersonic shock waves will send the aircraft out of control.*
* *Reduce the airspeed to 130 knots before landing. The speed and weight of the Learjet make it very difficult to stop, even with brakes, spoilers, and thrust reversers at work.*

This flight will be performed with the Learjet. Although we've flown the Learjet before, we didn't discuss its characteristics. So, for this flight, I will spend a little more time discussing how the Learjet handles. The Learjet cruises at around 418 knots. At this speed, the turns take a little longer, but the controls are more sensitive than in the Cessna. Keeping this in mind, let's get started.

Apply full throttle for the takeoff, and rotate at around 130 knots. The Learjet will want to climb at a steep angle, but don't let the nose get so high that you begin to lose airspeed. You'll notice on the artificial horizon instrument (also called the attitude indicator) a set of horizontal lines above and below the center triangle. A good climb angle for the Learjet is indicated when the top of the center triangle touches the top horizontal marker line. Maintain this climb angle until you reach an altitude of 8000 feet. Level off at 8000 feet, with the original heading of 264°. Reduce the power to a fan setting of 75 percent. After the aircraft has stabilized in straight and level flight, the airspeed should be about 420 knots.

Take Off in the Learjet

1. Press . (period) to disengage the parking brakes, if they are on.
2. Hold down keypad 9 to apply full throttle.
3. Rotate at 130 knots and press G to raise the gear after liftoff.
4. Keep the top of the center triangle in the artificial horizon instrument touching the top horizontal marker line while you climb to an altitude of 8000 feet.
5. At 7500 feet, use keypad 3 to reduce the power to a fan setting of 75 percent.
6. Level off at 8000 feet.

Okay, we're at 8000 feet, so let's make a right standard-rate turn to a heading of 60°. Doing this will head us out over Puget Sound toward Seattle. Because the Learjet does not have a turn coordinator, we'll use the artificial horizon instrument to indicate our turn. Below the horizon on the instrument, you will see a set of angled lines. As we make our standard-rate right turn, the right tip of the center triangle should line up with the first angled line.

Adventures in Flight Simulator

You will also notice that the turn seems to be taking a long time. This is due to the high speed of the Learjet. You might feel the urge to steepen the bank angle in order to get to the desired heading sooner. Don't! Instead, start the good habit of controlling the Learjet with slow, smooth movements. Level off when we reach a heading of 60°. Up ahead in the distance we should be able to see a patch of gray against the surrounding green scenery. That is Seattle. As we get closer, of course, we will be able to see the buildings.

Perform a Standard-Rate Right Turn to 60°

1. Apply right pressure to the yoke, lining up the right tip of the center triangle on the artificial horizon instrument with the first angled line, and make a standard-rate turn toward a heading of 60°.
2. Apply left pressure to the yoke at 50° to slowly level off at 60°, and return to straight and level flight.

Open the Nav/Com menu and select Autopilot. The only features we'll need for this flight are the altitude and wing leveler controls. Set the altitude for 8000 feet and turn on the wing leveler. Now the Learjet will maintain a constant altitude of 8000 feet and the wings will remain level, which will give us a chance to examine some of the different graphics settings as we fly. Let's also pause the simulator while we review some of the parameters for the various graphics modes.

Set Autopilot Options

1. Open the Nav/Com menu.
   Select Autopilot.
2. Set ALT (altitude) Hold to **8000**.
   Select LVL (wing leveler).
   Select Connected On from the Autopilot Switch box.
3. Press P to pause the simulator.

Graphics

Three different graphics modes are available with Flight Simulator 5: EGA, VGA, and SVGA. EGA supports 16 colors at a resolution of 640 x 350 pixels; VGA supports 256 colors at a resolution of 320 x 400 pixels; and SVGA supports 256 colors at a resolution of 640 x 400 pixels. SVGA is the preferred mode because

Cessna instrument panel display as it appears in SVGA (top) and VGA (bottom)

it offers the highest quality of graphics, but not everyone is fortunate enough to have SVGA. Although VGA works fine with Flight Simulator, SVGA is quickly becoming the video standard on 386 and 486 computer systems.

The most noticeable differences between VGA and SVGA are in the display of the Cessna cockpit and in the screen refresh rate (how frequently the pixels are updated). With its lower resolution, VGA has a better refresh rate because there are fewer screen pixels to update. However, SVGA displays the Cessna engine instruments next to the radios without your having to press Tab. The difference in the quality of other VGA and SVGA scenery is minimal, with the VGA graphics being a little more grainy. These differences are most noticeable in the details of the city buildings and the instrument panel display.

Display Preferences

The display preferences set in the current graphics mode can have a lot to do with screen refresh performance. The display parameters can be found in the Preferences selection on the Options menu. Although each of the settings can be desirable and can contribute to the realism of Flight Simulator, each adds a small performance cost. The available display preferences are as follows:

- **Image Quality/Speed** The options under this setting are High/Slow, Medium/Medium, and Low/Fast. The higher the image quality, the slower the refresh rate.

- **Flicker/Speed** Some video cards display a noticeable flicker at high video speeds. Changing these settings should help reduce or eliminate any flicker that occurs. The options are Much Flicker/Fast, Some Flicker/Medium, and No Flicker/Slow.

- **Text Presentation** This setting affects the display of the text information "issued" by air traffic control (ATC). The options are Single Line or Continuous, meaning that the information can be displayed one line at a time or as continuous scrolling text. Displaying one line at a time burdens performance less and does not slow down the display refresh rate. Continuous is the default.

- **Map Display** This setting affects how the map is coordinated in relation to the heading of the aircraft. The options are North Oriented, Aircraft Oriented, and North at High Altitude. Aircraft Oriented is the default. When North Oriented is selected, the map is always shown with north at the top. When Aircraft Oriented is selected, the map rotates around the center point (of the aircraft) as the aircraft changes position. And finally, when North at High Altitude is selected, the map rotates around the center point of the aircraft at lower altitudes, but at high altitudes (when ground details are not visible), north is always at the top of the map display.

- **Location Readout** This setting specifies which windows will display the location coordinates when Slew Mode is turned on.

- **SVGA Board Maker** If the current display is SVGA, this option lets you select the type of video card you use.

The following display preferences are check-box toggles that turn features on or off.

- **See Ground Scenery Shadows** When this option is turned on, you can see shadows from buildings.

- **See Aircraft Shadows** When this option is turned on, you can see the shadow of the aircraft.

- **See Propeller** When this option is turned on, Spot views show a propeller that changes its rate of rotation depending on engine speed.

- **Landing Lights Available** This option supplies the very bright landing lights on the aircraft that are mostly used for seeing the runway and taxiway surfaces at night.

- **See Own Aircraft From Cockpit** When this option is turned on, the various internal views also show the interior of the aircraft. This adds realism, but can also restrict the view out the back or right side, especially in the Learjet.

- **Textured Sky** When this option is turned on, the clouds look real instead of like large polygons. This option is well worth keeping on all the time.

- **Gradient Horizon** This option gives a more realistic look to the distant horizon and is also worth keeping on.

- **Smooth Transition View** This option affects the smoothness of the transition when you switch between the various views, such as when you switch from Cockpit view to Spot view. This option is on by default.

- **Textured Buildings** This option provides a smoother, more realistic view of city buildings.

- **Textured Ground** When this option is turned on, the ground looks very realistic from the air. Close up, however, the ground looks grainy and rough.

- **Aircraft Texture** When this option is turned on, the external appearance of the aircraft is much more detailed.

Settings for Optimum Performance

To see the impact on performance when these features are on, let's return to our Learjet flight and change some of the settings. Before unpausing the simulator, enter the following changes to the display parameters:

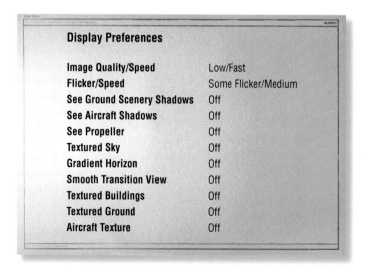

Display Preferences	
Image Quality/Speed	Low/Fast
Flicker/Speed	Some Flicker/Medium
See Ground Scenery Shadows	Off
See Aircraft Shadows	Off
See Propeller	Off
Textured Sky	Off
Gradient Horizon	Off
Smooth Transition View	Off
Textured Buildings	Off
Textured Ground	Off
Aircraft Texture	Off

Change the Display Preferences

1. Open the Options menu.
   Select Preferences.
   Select Display.

2. In the Image Quality/Speed box, select Low/Fast.
   In the Flicker/Speed box, select Some Flicker/Medium.

3. Be sure that See Ground Scenery Shadows, See Aircraft Shadows, See Propeller, Textured Sky, Gradient Horizon, Smooth Transition View, Textured Buildings, Textured Ground, and Aircraft Texture are all unselected (turned off).

4. Press P to unpause the simulator.

A steep turn is the best situation for visually testing the display refresh rate. Let's first turn off the Autopilot and then enter a steep left turn to a heading of 180°. We'll use the artificial horizon instrument for determining the angle of bank. When the third mark on the top left side of the instrument lines up with the top pointer, we're in a steep turn. While we're in the turn, notice the speed of the refresh rate, which is evident in the smoothness of the display. The higher the refresh rate, the smoother the display. When we reach a heading of 180°, return to straight and level flight, turn on Autopilot again, and pause the simulator once more.

Chapter 5: Graphics, Scenery, Pictures, and Video

Make a Steep Left Turn to 180°

1. Press Z to turn off the Autopilot.
2. Apply left pressure to the yoke and line up the top pointer of the artificial horizon instrument with the third mark on the top left of the instrument to make a steep left turn to a heading of 180°.
3. Notice the display refresh rate during the turn.
4. Apply right pressure to the yoke at 190° to slowly level off at 180° and return to straight and level flight.
5. Press Z to turn on the Autopilot.
6. Press P to pause the simulator.

Settings for the Best Graphics

Now that we have seen the display refresh rate when the display parameters are set for optimum performance, let's change the settings to get the best graphics. Use the settings shown in this table:

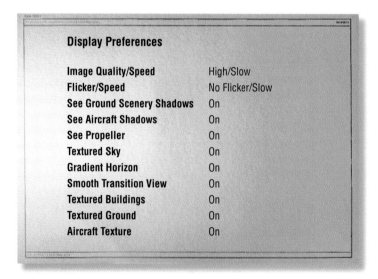

Display Preferences	
Image Quality/Speed	High/Slow
Flicker/Speed	No Flicker/Slow
See Ground Scenery Shadows	On
See Aircraft Shadows	On
See Propeller	On
Textured Sky	On
Gradient Horizon	On
Smooth Transition View	On
Textured Buildings	On
Textured Ground	On
Aircraft Texture	On

Change the Display Preferences

1. Open the Options menu.
   Select Preferences.
   Select Display.
2. Set Image Quality/Speed to High/Slow.
   Set Flicker/Speed to No Flicker/Slow.
3. Select See Ground Scenery Shadows, See Aircraft Shadows, See Propeller, Textured Sky, Gradient Horizon, Smooth Transition View, Textured Buildings, Textured Ground, and Aircraft Texture.
4. Press P to unpause the simulator.

Again, we'll test the display refresh rate by making a steep turn. Begin by turning off the Autopilot, and then make a steep right turn to a heading of 60°. Notice how much slower the refresh rate is. If we were close to the ground and turning past some buildings, the refresh rate would be even slower. After you reach a heading of 60°, level off and turn the Autopilot back on. Then pause the simulator again.

Make a Steep Right Turn to 60°

1. Press Z to turn off the Autopilot.
2. Apply right pressure to the yoke to line up the top pointer of the artificial horizon instrument with the third mark on the top right of the instrument to make a steep right turn to a heading of 60°.
3. Notice the display refresh rate during the turn.
4. Apply left pressure to the yoke at 50° to slowly level off at 60° and return to straight and level flight.
5. Press Z to turn on the Autopilot.
6. Press P to pause the simulator.

Multiple Views

When real pilots use Flight Simulator, one of their biggest complaints is the lack of peripheral view. For example, when flying a tight approach pattern and landing, it is difficult to determine where the runway is and what the attitude of the aircraft is in relation to the runway. In Flight Simulator, you can look in many directions, but in any particular direction you can only see straight ahead. Consequently, the best you can do is to keep switching among the Cockpit views and scrolling around the aircraft in Spot view to keep an eye on the position of the runway. One way around this problem is to use the View 2 window and the Map Display window.

Wing Tip:
Depending on the type of computer you are using, the display refresh rate might or might not be drastically affected by changes to the display parameters. On a 386 DX running at 33MHz, the changes are more noticeable than on a 486 DX running at 66MHz. If you are running a 486 at 66MHz, you might not notice much of a difference in the refresh rate when you perform steep turns.

Having View 1, View 2, and the Map Display on the screen at the same time gives you a lot more information about the position and attitude of the aircraft and helps compensate for the lack of a true peripheral view.

To see how this works, let's go back to our Learjet flight and add a few windows. Before unpausing the simulator, you need to resize the View 1 window (the current view, from the cockpit) to make room for the other windows. In the Cessna or the Sopwith Camel, you can do this with the mouse or the menus. But because we are using the Learjet, the lower right corner of the View 1 window is covered by the instrument panel, and you cannot access it with the mouse. Therefore, the sizing and moving will have to be done with menu selections. Use the Views menu and

Adventures in Flight Simulator

the Size And Move Windows command to resize View 1. Keep the height of the View 1 window the same, but bring the right side of the window to the left until it is at the center of the display. The View 1 screen should now be covering only the left half of the display above the instrument panel. Now we can turn on the other views. From the Views menu, select View 2 and then select Map View. View 2 will be Spot view, and the third window will display Map view.

Turn on Multiple Views

1. Open the Views menu.
   Select Size And Move Windows.
   Select View 1.

2. Hold down the Shift key and use keypad 4 to change the size of the View 1 window.
   Move the right side of the window to the center of the display screen.

3. Press Enter to return to the Size And Move Windows dialog box, and click OK.

4. Open the Views menu.
   Select View 2.

5. Open the Views menu again.
   Select Map View.

6. Open the Views menu again.
   Select Size And Move Windows.
   Select View 1.

7. Hold down the Shift key and use keypad 6 to resize the View 1 window to meet the
   left edge of the View 2 and Map view windows, and press Enter twice.

You will notice that using multiple views makes it a lot easier to determine the attitude and position of the aircraft. The drawbacks are that the refresh rate is slightly reduced and that using multiple windows is less realistic.

You can use the zoom controls on the active window (the window with the thin white border around it), and you can make any of the three view windows active by clicking the mouse in it. To get a better impression of where we are in the Learjet, make the Map view active and zoom out until you can clearly see the Sound west of Seattle. Let's also look at the Learjet from various external views. Select the View 2 window, and then pan around the aircraft, ending at the rear of the aircraft.

Zoom the Map View and Pan in Spot View

1. Press P to unpause the simulator.
2. Click the mouse in the Map view to make it the active window.
   (When a window is active, its border is highlighted.)
3. Press the – (minus) key at the top of the keyboard several times to zoom out until you can clearly see Puget Sound.
4. Click the mouse in the View 2 window to make it the active window.
5. Press Shift and any of the keypad keys to pan completely around the aircraft.
6. Press Shift-keypad 2 to end the pan at the rear of the aircraft.
7. Press P to pause the simulator.

Scenery

You can also adjust the scenery in Flight Simulator 5 to your preference and tune it for performance. The same trade-offs that applied to the graphics parameters apply to the scenery. The better the detail, the more impact it has on the display refresh rate. The options for tuning the scenery system are on the Scenery menu, under Scenery Complexity. You can set the parameters for both View 1 and View 2. The available parameters are as follows:

- **Stars In the Sky**   This option displays the stars in the evening, at night, and in the early morning.
- **Approach Lighting**   This option supplies the approach lights to an airport.
- **Horizon Only (No Scenery)**   As the name implies, when this option is turned on, no scenery is displayed.
- **Wire Frame Polygons**   Applies only to Flight Simulator 4.0 scenery.
- **Image Complexity**   This allows the setting of the density of the scenery. The available settings are Very Sparse, Sparse, Normal, Dense, and Very Dense. Very Dense is the best, but it slows down the display refresh rate.
- **Earth Pattern**   Applies only to Flight Simulator 4.0 scenery.
- **Moonlight At Night (View 1 and 2)**   This option displays moonlight at night for Views 1 and 2.
- **Image Smoothing (View 1 and 2)**   This option reduces graininess but creates a less clear display for Views 1 and 2.

Maximum Scenery

Let's see how the settings for scenery will affect our Learjet flight. For starters, use the settings shown in the table:

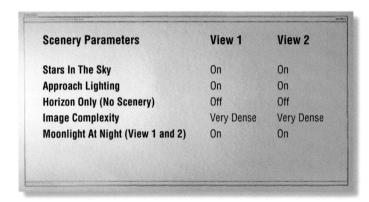

Scenery Parameters	View 1	View 2
Stars In The Sky	On	On
Approach Lighting	On	On
Horizon Only (No Scenery)	Off	Off
Image Complexity	Very Dense	Very Dense
Moonlight At Night (View 1 and 2)	On	On

Set the Scenery Parameters

1. Open the Scenery menu.
   Select Scenery Complexity.

2. Select View 1, and be sure that Stars In The Sky and Approach Lighting are selected and that Horizon Only (No Scenery) is not selected.
   Set Image Complexity to Very Dense.

3. Select View 2, and set the same parameters you set for View 1.

4. Select Moonlight At Night (View 1 and 2).

5. Press P to unpause the simulator.

Now that we are flying again, look for Seattle. If the aircraft is not facing toward Seattle, make a gentle turn to the correct heading. The city looks really beautiful in the evening, so fly in a little closer for a good look. We also need to change the time to make the sky a little darker to get the full effect of the stars and lights. Advance the time to 17 hours and 15 minutes (5:15 p.m.).

Wing Tip:
You can greatly increase the visual impact of the nighttime scenery by dimming the room lights.

Fly to Seattle and Set the Time

1. Use the yoke to turn toward Seattle.
2. Open the World menu.
   Select Set Time and Season.
   Select Set Exact Time.
   Set Hours to **17** and Minutes to **15**.

Notice how the city becomes more visible at night. We can also see the stars between the pink clouds. Now that the center of the city is more visible, we can adjust the direction of the aircraft to head for downtown Seattle. To get a good look, we also need to reduce speed and lower the altitude. Reduce the speed to a fan setting of 50 percent, turn off the Autopilot, and then descend to an altitude of 3000 feet. After we reach 3000 feet, set the Autopilot altitude lock to 3000 feet and turn the Autopilot back on.

Reduce Speed and Altitude

1. Use the yoke to direct the aircraft so that it will pass over the center of Seattle.
2. Use keypad 3 to reduce the throttle to a fan setting of 50 percent.
3. Press Z to turn off the Autopilot.
4. Apply slight downward pressure on the yoke to descend to 3000 feet.
5. Open the Nav/Com menu.
   Select Autopilot.
   Set ALT (altitude) Hold to **3000**.
6. Press Z to turn the Autopilot back on.

At about 17:18 (5:18 p.m.), the red cockpit lights will come on. As we draw closer to Seattle, it would be a good idea to zoom in the Map view a little to display more detail of the city.

Minimum Scenery

At this point, we can test the effects of setting the scenery parameters for a minimum of scenery. Pause the simulator and use the settings shown in the table:

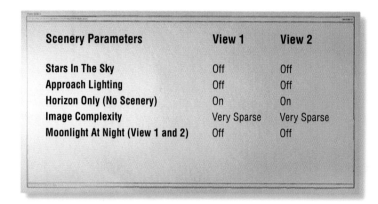

Scenery Parameters	View 1	View 2
Stars In The Sky	Off	Off
Approach Lighting	Off	Off
Horizon Only (No Scenery)	On	On
Image Complexity	Very Sparse	Very Sparse
Moonlight At Night (View 1 and 2)	Off	Off

Turn Off the Scenery

1. Press P to pause the simulator.

2. Open the Scenery menu.
   Select Scenery Complexity.

3. Select View 1.
   Deselect (turn off) Stars In The Sky and Approach Lighting.
   Select (turn on) Horizon Only (No Scenery).
   Select Very Sparse in the Image Complexity box.

4. Select View 2, and set the same parameters as you set for View 1.

5. Deselect Moonlight At Night (View 1 and 2).

6. Press P to unpause the simulator.

The scenery of the city, the lights, the roads, the water, and the stars is gone. All we have now are checkerboard patterns on the ground and clouds in the sky. Although such minimal scenery can increase the refresh rate, it can also decrease your enjoyment of the program. So let's turn the scenery back on, turn off the Autopilot, drop the altitude to 1000 feet, and then skim over the bright lights of Seattle at a whopping speed of 340 knots.

Turn On the Scenery and Buzz Seattle

1. Press P to pause the simulator.

2. Open the Scenery menu.
   Select Scenery Complexity.
   For View 1, select Stars In The Sky and Approach Lighting.
   Deselect Horizon Only (No Scenery).
   Select Very Dense in the Image Complexity box. Select the same settings for View 2.
   Select Moonlight At Night (View 1 and 2).

3. Press Z to turn off the Autopilot.

4. Apply downward pressure to the yoke, and descend to 1000 feet.

5. Fly over the center of Seattle.

Now that is fun! Thank goodness it was dark. I would hate to see us get turned in for reckless flying! If you want to do some more buzzing of Seattle, maybe getting a close-up view of the Space Needle, feel free to continue flying. When you are done, we will discuss the features of dynamic scenery.

Dynamic Scenery

Dynamic scenery contributes to realism by adding boats, aircraft, and moving ground traffic. Most of the dynamic scenery is in the Chicago area at Meigs Field and O'Hare. We can take a look at some of this scenery by loading and running the default startup situation that comes with Flight Simulator 5. Load FS5.STN, the situation named Meigs Takeoff Runway 36.

Load Meigs Takeoff Situation

1. Open the Options menu.
   Select Situations.

2. Select Meigs Takeoff Runway 36, filename FS5.STN.

The dynamic scenery settings are as follows:

- **Air Traffic**   This option toggles aircraft traffic in the sky on or off. Other aircraft usually appear as moving dots in the distance until you get closer.
- **Aircraft Ground Traffic**   When this option is turned on, aircraft taxi around in the airport, take off, and land. It is a good idea to check for approaching aircraft before moving onto the runway when this option is turned on.
- **Airport Service Traffic**   Ever have an encounter with a fuel truck? It is possible when this option is turned on.
- **Traffic Outside Airports**   You'll encounter boats, hot air balloons, and other types of traffic outside the airport areas when this is turned on.
- **Scenery Frequency**   The settings for this option range from Very Sparse, for a minimum amount of traffic, to Very Dense, for the maximum amount.

Use the following settings to demonstrate dynamic scenery:

Dynamic Scenery Settings	View 1
Air Traffic	On
Aircraft Ground Traffic	On
Airport Service Traffic	On
Traffic Outside Airports	On
Scenery Frequency	Very Dense

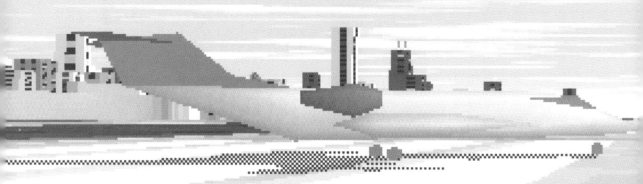

Set Dynamic Scenery

1. Open the Scenery menu.
   Select Dynamic Scenery.

2. Select View 1.
   Select Air Traffic, Aircraft Ground Traffic, Airport Service Traffic, and Traffic Outside Airports.
   Be sure that Scenery Frequency is set to Very Dense.

Now all we have to do is wait and watch. After a minute or so, we will see an aircraft taxi in front of us and take off. As you can see, by the time it reaches the end of the runway and takes off, it becomes a tiny dot. Usually the first plane off is a Cessna, followed shortly by a Learjet. Now let's try something fun and see if we can take off directly behind another Cessna and follow it around the sky.

Follow the Leader

Flight Plan: For this flight, we'll use the Cessna Skylane R182 and will attempt to fly in formation with another Cessna taking off from Meigs Field.

✈✈✈

First of all, you'll need to save the current situation. This flight is difficult, so it will take a few attempts to be able to fly in formation with one of the dynamic scenery Cessnas. We have already watched a Cessna and a Learjet go by. When another Cessna pulls out onto the runway, get ready to apply full throttle. Wait until the other Cessna is facing down the runway, and then apply full throttle. You'll need a lot of control with the throttle and the yoke. The last thing you want to do is run into the other aircraft. Try to keep the other Cessna off to your left and ahead. Leave enough space to allow yourself time to react to the movements of the other aircraft.

Take Off in Formation

1. Open the Options menu.
   Select Save Situation, and save this situation as **CH5FLT2**. You can use **Follow the Leader** as a description.
2. Wait for another Cessna to pull onto the runway and line up for takeoff.
3. Use keypad 9 to apply full throttle and take off in formation with the other Cessna.
4. Using the yoke and keypad throttle controls, attempt to keep the other Cessna in front of you and to the left.
5. Leave enough space to allow you time to react to the movements of the other Cessna.

The longer you can follow the other aircraft, the better your piloting skills. Because most of the dynamic scenery aircraft here are flying the pattern, staying with the other aircraft until it lands again at Meigs is the ultimate test of flying skill. Sound like a challenge? I think it can be done....

Pictures

What better way to prove your skill at flying than with a picture? The new Flight Photograph feature of Flight Simulator 5 is a great way to hold onto fond memories of those white-knuckle flights. It is also a good way to brag about your exploits by sharing the pictures with your friends or uploading them to the flight simulation group on CompuServe, Genie, or any other bulletin board or information service where armchair pilots hang out.

To demonstrate Flight Photograph, let's restart the last situation (Follow the Leader) and take a flight photograph right after taking off behind the other aircraft. This time, switch to Spot view from behind the aircraft, and lower the view to 5 feet. As soon as we lift off the runway, open the Views menu and select Flight Photograph. Enter a name for the picture, and then click OK. You will notice that Flight Simulator "freezes" for a few seconds while the photograph is being stored to disk.

Take a Flight Photograph

1. Open the Options menu.
   Select Situations.
   Select CH5FLT2.STN, and restart the Follow the Leader situation.

2. Press S twice to go to Spot view, and press Shift-keypad 2 to position the view from behind the aircraft.

3. Open the Views menu.
   Select Set Spot Plane.
   Set Altitude to **5**.

4. Wait for a Cessna to pull onto the runway, and then take off right behind it.

5. After lifting off the runway, open the Views menu.
   Select Flight Photograph, and type a name in the Filename box.

6. Press P to pause the simulator.

That's all there is to it. The file is saved in a PCX format, and you can view it with any drawing or picture viewing program that supports PCX files, like Microsoft Paintbrush running under Microsoft Windows.

Video

So you think pictures are not enough to convey the true action of the moment? Well, we could set up a camcorder on a tripod and point it at the monitor while we fly. Or we could use the built-in video recorder in Flight Simulator.

With the video recorder, we can record and play those moments that only video can capture. Let's give it a try by continuing our last flight (from just after taking the flight photograph), and we'll record a great video while buzzing Chicago.

Before unpausing the simulator, open the Options menu and select Video Recorder. Click the Record New Video button. The next selection box will show the default Recording Interval—one second—and will tell us to press the Backslash key to stop recording.

Click OK to start recording. Text will appear on the screen to let us know that we are currently recording at one-second intervals. Turn the aircraft toward Chicago, and fly past a few buildings at a low level. After passing through downtown Chicago, press the Backslash key to stop the recording. Another box will pop up and ask for a description and a filename. Enter **Buzzed Chicago** as the description. BUZZEDCH will show up as the filename. You can change the filename if you want. Then click the Save Video button to save the video to disk.

Record a Video

1. Open the Options menu.
   Select Video Recorder, and click the Record New Video button.
2. Click OK to accept the default recording interval of one second.
3. Turn the aircraft toward Chicago, and buzz a few buildings.
4. Press the Backslash key to stop the recording.
5. Type **Buzzed Chicago** in the Video Title box, accept BUZZEDCH in the Filename box, and then click the Save Video button to save the video.

Now let's load up the video recorder again and play the video we just saved. This time, select the video named BUZZEDCH and click the Play Selected Video button. Another message box will tell you to press the Esc key when you have finished viewing the video. You can also put the video into a constant loop by clicking on the check box next to Repeat Replay. Go ahead and click on the check box, and then click OK.

Play a Video

1. Open the Options menu, and select Video Recorder.
2. Select the video named BUZZEDCH, click the Play Selected Video button, click on the Repeat Replay check box, and then click OK.

In this chapter, we covered the following topics:

Flying the Learjet
Using Autopilot with altitude hold and wing leveler
Changing the modes and observing the effect on performance
Changing the display preferences and observing the effect on performance
Setting up and using multiple views
Changing the scenery parameters and observing the effect on performance
Changing the dynamic scenery parameters and observing the effect on performance
Flying in formation with dynamic scenery
Taking and viewing flight photographs
Recording and viewing videos

ADVENTURES 1-6

ADVENTURE no. 1

RADIO-CONTROLLED FLIGHT

Now that you are a proficient Flight Simulator pilot, it's time to exercise your flying skills in various adventures that will determine whether you have the "right stuff." In this first adventure, you'll simulate flying a radio-controlled aircraft—without having to build the plane first!

Flight Plan: For this adventure, we'll fly the Cessna Skylane RG R182 out of Meigs Field. The keyboard or joystick will serve as the "radio." From the vantage point of Tower and Spot views, we'll fly the radio-controlled Cessna over Chicago and attempt a landing at the airstrip. Then we'll take off in the Sopwith Camel to fly some other maneuvers.

You can load most of the settings that we need from the standard Meigs situation, Meigs Takeoff Runway 36, which comes with Flight Simulator. The following settings are in addition to the default settings in the standard situation. Therefore, after you load the situation, you don't need to enter location, weather, or any settings other than the ones mentioned in the Flight Settings table for this flight.

After you enter the flight settings, View 1 shows only gray. This is okay. Because you set Tower view to the same location as the aircraft and set the zoom factor to 32 (magnified 32 times), all you can see is a close-up of the aircraft fuselage. When the aircraft starts to move down the runway, the view of the aircraft will change.

Flight Settings

Situation Name	Meigs Takeoff Runway 36
View Options	
View 1	On, Tower
View 1 Zoom	32.00
View 2	On, Spot
View 2 Zoom	1.00
Titles on Windows	On
Set Exact Location	Set Tower view from aircraft location
Auto Coordination	On

Enter the Flight Settings

1. Open the Options menu.
   Select Situations.
   Load the situation called Meigs Takeoff Runway 36.

2. Open the World menu.
   Select Set Exact Location.
   Choose Set Tower View (From Aircraft Location).

Adventure 1: Radio-Controlled Flight

Aeronautically Speaking: Normally, in real-world radio-controlled flight and in real aircraft flight, the pilot always tries to take off facing into the wind. Because lift is generated by the amount of air passing over the wings, heading into the wind allows the aircraft to use a shorter runway for a quicker takeoff. The same applies to landing the aircraft; the pilot always tries to land into the wind.

3. Open the Views menu.
   Select View Options.
   Select View 1, set Status to On, set View to Tower, and set Zoom to **32**.
   Select View 2, set Status to On, set View to Spot, and set Zoom to **1**.

4. Open the Views menu again.
   Select Size And Move Windows, and follow the displayed instructions for sizing and moving windows. Size the Tower view (it is totally gray) for the upper left half of the view screen, and size the Spot view for the upper right half of the view screen.

5. Open the Sim menu and make sure that Auto Coordination is checked.

6. Click once on the left window to make it the active window.
   (When a window is active, its border is highlighted.)

7. Open the Options menu and save the situation as **ADV1AFLT**.

Radio On, Aircraft On, Take Off

Actually, we are cheating a little because when you fly a radio-controlled aircraft, you do not have a spot-plane perspective, nor can you see an instrument panel. But because we will attempt to land on the airstrip and we don't have the advantage of panoramic vision, we'll use the extra visual input from these views to help us land. In the second part of the adventure, we'll change the view for a more accurate simulation of radio-controlled flight.

Imagine standing over a miniature Cessna R182. Apply full throttle and take off. As the aircraft gets farther away, increase the zoom setting for a closer view of it. (Yes, we're cheating again. Just think of it as having a good pair of binoculars!) Climb to 2000 feet, and then make a left turn to a heading of 180°.

Take Off

1. Press . (period) to disengage the parking brakes if they are on.

2. Use keypad 9 to apply full throttle, and rotate at 80 knots.

3. Use the + (plus) and – (minus) keys at the top of the keyboard to zoom the view as necessary.

4. Press G to raise the landing gear after liftoff, and climb to 2000 feet.

5. Use keypad 3 to reduce the throttle to 80 percent, and make a standard-rate left turn to a heading of 180°.

Maintain an altitude of 2000 feet, and stay at a heading of 180°. Use the zoom as needed for maintaining visual contact with the aircraft. (Many a radio-control pilot has lost a plane by looking away for a second.) Keep an eye on the aircraft as it flies past our vantage point on the ground. When the left wing of the aircraft is lined up with our view position at the end of the runway, the aircraft will be directly west of us. Because we are on the end of the runway, this is a good marking point to begin our approach. Begin by reducing the throttle 50 percent. Apply full flaps, and lower the landing gear.

Wing Tip:
It would be a good idea to save this current position as a situation. Because landing using only views from outside the aircraft is a little tricky, you might want to practice a few times, starting at this point.

Complete the Downwind Leg

1. Fly straight and level at 2000 feet at a heading of 180°.
2. As the aircraft comes directly west of our position, use keypad 3 to reduce the throttle to 50 percent, press F8 to apply full flaps, and press G to lower the gear.

From this point, we need to extend the downwind leg a little so that we have enough distance from the airstrip to make a good approach. Continue with the heading of 180° for another 15 seconds. Then turn left to a heading of 000°. When the Cessna comes out of the turn, it should be lined up with the airstrip, or fairly close to it. Use the Spot view on the right side of the screen to check the alignment. Reduce the throttle to full off.

Complete the Base Leg

1. Continue the downwind leg at a heading of 180° for another 15 seconds.
2. Turn left to a heading of 000°.
3. Use Spot view to check alignment with the airstrip.
4. Use keypad 3 to reduce the throttle to full off.

From here on, it's just a matter of keeping the aircraft lined up with the airstrip and adjusting the power as necessary to maintain a proper descent. The aircraft is now flying toward us, so you will need to adjust the zoom for an appropriate view of it. Try to have the aircraft as level as possible when you pass the threshold markers, and then gently flare for the landing. Don't worry if it's a little rough. As soon as the aircraft touches down completely (or has stopped bouncing), turn off the throttle and at 50 knots apply the brakes for a full stop.

Adventure 1: Radio-Controlled Flight

Land the Aircraft

1. Maintain a straight approach to the airstrip and add a little throttle if necessary.
2. Adjust the zoom as needed.
3. When you reach the end of the runway, gently flare for the landing.
4. At touchdown, reduce the throttle to full off, and at 50 knots press . (period) to apply the brakes and stop.

Loop the Loop

To spice up our next radio-controlled maneuver, the inside loop, we'll use the Sopwith Camel. Although this little airplane doesn't have the sophisticated controls that the Cessna does, it is very maneuverable and can be fun to fly.

Reset the ADV1AFLT.STN flight situation, change your aircraft to the Sopwith Camel, and then save this situation as ADV1BFLT. Now your large screen should be green, the fuselage color of the Camel. (Flight Simulator does not save window resizings when you save a situation, so when you reset ADV1AFLT.STN you will see the default sizes of View 1 and View 2. Don't worry about this now; you'll only be using Tower view for this flight anyway.) Apply full power and rotate at 70 knots. At this point, we can dispense with the extra views and try some more realistic radio-controlled flying. After you are in the air, maximize the Tower view by pressing the W key. You can still use the zoom to keep an eye on the

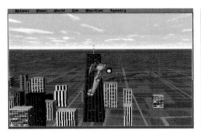

aircraft. Continue to climb for about a minute to give the aircraft some altitude. Because you no longer have the advantage of an instrument panel or Spot view, you'll have to rely on your own judgment to determine whether the aircraft is at the proper altitude. (Or you can always sneak in a W-key toggle if you lose faith in your judgment!)

Enter the Flight Settings and Take Off

1. Open the Options menu, and select Reset Situation to reload the ADV1AFLT.STN situation.
2. Open the Options menu again, select Aircraft, and then select Sopwith Camel.
3. Open the Options menu again, and save the situation as **ADV1BFLT**.
4. Press . (period) to disengage the parking brakes if they are on.
5. Use keypad 9 to apply full power, and rotate at 70 knots.
6. Press W to maximize the Tower view, and continue climbing for about a minute.

When you think the aircraft has enough altitude, turn it left until the aircraft is headed in your direction so that it will fly directly over your position. The next maneuver will be an inside loop. To prepare for this, after the aircraft comes out of the turn, point the nose down toward you. This will let the aircraft build up enough speed for the loop. When you think the aircraft has enough airspeed, gently pull back on the yoke and watch as the aircraft goes through a full inside loop. Remember that when the aircraft is coming out of the loop it has a lot of airspeed and will tend to start flying another loop. To counter this tendency, apply forward pressure on the yoke before the nose comes up level with the horizon.

Perform an Inside Loop

1. Turn left until the aircraft is heading toward you.
2. Point the nose of the aircraft down to descend and build up airspeed.
3. When you have enough airspeed, pull back on the yoke and perform an inside loop.
4. Level off after the loop.

Wing Tip:
Remember, when you are controlling an aircraft that is coming toward you, the controls work in reverse. Moving the yoke to the left causes the aircraft to move toward your right, and vice-versa.

Other Maneuvers

Another neat trick to perform with a radio-controlled aircraft is to do a low-flying pass. Radio-control pilots just love to buzz themselves! To do this, you start by pointing the nose of the aircraft toward you.

As the aircraft gets closer, gently control the altitude to keep the aircraft just above the ground as it flies past. This might be a good time for a flight photograph!

Buzz the Pilot

1. Adjust the aircraft's heading so that it is coming toward you again.
2. Point the nose down to descend.
3. As the aircraft gets closer to the ground, level off without touching down.
4. Keep the aircraft slightly above the ground and headed toward you.
5. Open the Views menu and take a flight photograph.

Now that you know how to control the aircraft, feel free to continue flying and do more stunts and maneuvers. A good source of aerobatic maneuvers is Chapter 14, "Aerobatics Course," in the *Flight Simulator Pilot's Handbook*.

If you want to try a more realistic approach to flying radio-controlled aircraft, fly this adventure again with the zoom at a constant setting of 32. When the aircraft is almost too far away to see, turn it toward you again. If you decide to try landing, press W to use the Spot view.

If you are feeling really brave, try an inverted fly-by! To do this, perform another inside loop, but this time, when the aircraft is upside down and at the top of the loop, keep it in the inverted position and don't finish the loop. Remember that when you're in inverted flight all the controls are reversed. You climb by pushing the yoke forward and descend by pulling back on the yoke. You make a left turn by moving the yoke to the right, and vice-versa. Flying upside down is tricky and will rapidly give you an appreciation of the stunt pilot's skill!

ADVENTURE no. 2

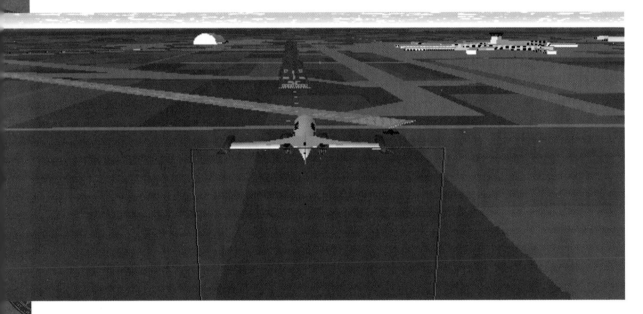

CORPORATE PILOT

It's 4:00 in the morning, and the phone startles you out of a dream. The president of your company needs to be in Chicago for a meeting at 9:00 a.m. A flight from New York–John F. Kennedy International Airport to Chicago–O'Hare takes about 2 hours and 15 minutes, so you get busy. As soon as you get dressed, you give the FSS (flight service station) a call to check on the weather. Just some scattered clouds at 10,000 feet—nothing to worry about. You then call the FBO (fixed base operator) and ask them to pull the Learjet out of the hanger, fill it with fuel, and put on a pot of coffee for the president and his staff. When you arrive at the airport at 4:40, you call the FSS again to double-check on the weather. There is no change, so you file the flight plan and enter the flight settings.

Flight Plan: For this flight, we'll depart in the Learjet from Kennedy International Airport at 5:00 a.m. and fly to Chicago–O'Hare Airport. We'll use the Autopilot and the EFIS system to make the flight easier.

Flight Settings	
Location	New York Kennedy Intl.—Runway 31L
Aircraft	Learjet 35A
Season	Autumn
Time of Day	5:00 a.m.
NAV Radio 1	115.9
OBI 1 Instrument	270°
NAV Radio 2	114.2
OBI 2 Instrument	280°

Enter the Flight Settings

1. Open the Options menu.
   Select Aircraft.
   Select Learjet 35A.

2. Open the World menu.
   Select Airports.
   Select USA-New York from the Choose A Scenery Area menu.
   Select New York Kennedy Intl.-Runway 31L from the Choose The Airport You Want To
      Fly From list box.

3. Open the World menu again.
   Select Set Time and Season.
   Set Season to Autumn.
   Select Set Exact Time, and type **5** in the Hours box and **0** in the Minutes box.

Adventure 2: Corporate Pilot

4. Open the Nav/Com menu.
   Select Navigation Radios.
   Set NAV 1 Frequency to **115.9** with an OBS Heading of **270**.
   Set NAV 2 Frequency to **114.2** with an OBS Heading of **280**.

5. Open the Options menu.
   Select Save Situation.
   Save the situation as **ADV2FLT**.

Departure

The president and his staff are on board, and we're sitting at the end of the runway waiting for clearance. I'm your co-pilot, and I'll take care of the radio work.

"JFK tower. Lear November Five Foxtrot Sierra Lima, holding on runway three one left."

"Lear Sierra Lima, you are cleared for takeoff. Contact departure control at two thousand feet."

"Roger, JFK tower. Sierra Lima taking off."

It's time for takeoff. Release the parking brakes, apply full throttle, and rotate at 190 knots for a smooth takeoff. As soon as the plane is in the air, apply some nose-down pressure to keep the climb from being too steep, and then raise the gear. Set the angle of climb with the artificial horizon instrument by placing the top white horizontal bar at the top of the pyramid.

Take Off

1. Press . (period) to release the parking brakes.
2. Hold down keypad 9 to apply full throttle.
3. Rotate at 190 knots.
4. Apply nose-down pressure to decrease the angle of climb.
5. Set the climb angle so that the top horizontal bar of the artificial horizon instrument is at the top of the pyramid. (This is a climb rate of approximately 6000 feet per minute.)

"Departure control, this is Lear November Five Foxtrot Sierra Lima, out of two thousand from runway three one left."

"Sierra Lima, continue climb to flight level three two and turn left to heading two seven zero. Contact New York Center."

"Roger, departure. Climbing to three two, heading two seven zero. Good day..."

"...New York Center, Lear November Five Foxtrot Sierra Lima, climbing out of seven thousand for flight level three two and turning left to two seven zero."

"Sierra Lima, continue as advised. Notify at flight level three two."

"Roger, Sierra Lima."

Make a left turn to a new heading of 270° and continue climbing. We'll pass through some clouds between 9500 feet and 11,500 feet. When we reach the target heading, level the wings but continue the climb. As we get above 16,000 feet, the rate of climb will begin to drop a little because the air is less dense. The nose will tend to come down a little, so apply a little back pressure to the yoke when necessary. When we get over 25,000 feet, decrease the angle of climb. Keep the horizon under the top of the instrument panel. At 28,000 feet we should be about 24 miles from the airport, according to the DME on the OBI. As we reach 31,600 feet, drop the nose and reduce the throttle to a fan setting of 80 percent. Level off at 32,000 feet.

Turn Left to 270°

1. Turn left to a new heading of 270° during your climb.
2. Continue to climb to an altitude of 32,000 feet.
3. At 32,000 feet, level off and use keypad 3 to reduce power to a fan setting of 80 percent.

Adventure 2: Corporate Pilot

Time for Autopilot

"New York Center, Lear Sierra Lima is level at flight level three two."

"Roger, maintain current course."

"Sierra Lima, roger."

Now that we are level at our target altitude and heading, we can let the Autopilot control the aircraft. Select the Autopilot, and turn on the altitude and heading holds. By now, we should be about 44 miles away from the New York airport. We'll lose contact with the VOR at about 75 miles.

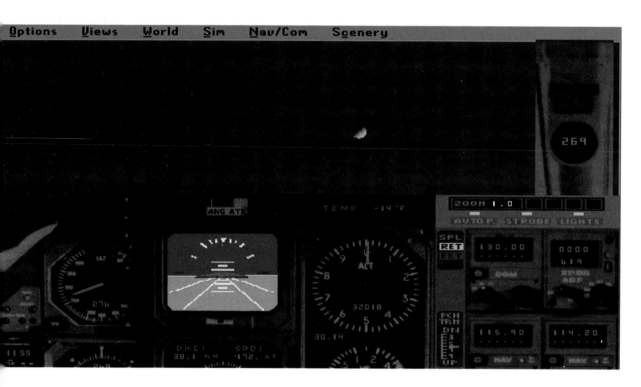

Turn On the Autopilot

1. Open the Nav/Com menu, and select Autopilot.
2. Set ALT (Altitude) Hold to **32000**.
3. Set HDG (Heading) Hold to **270**.
4. Set Autopilot Switch to Connected (On).

Once we've lost contact with the VOR, switch over to the OBI 2 instrument. The frequency dialed into the NAV 2 radio should be 114.2, the frequency for the Chicago Heights VOR, which is located southeast of Chicago. Flying to this VOR will allow us to prepare for our instrument approach into Chicago–O'Hare Airport.

Look out the left window and notice that the horizon is slanted down toward the right. This means that the nose of the aircraft is pointed up over the horizon even though we are flying straight and level. Because of the thin air at this altitude, in order to keep enough air flowing over the wing to provide lift for level flight, the wing needs to have a higher angle of attack.

Switch to Chicago Heights VOR

1. Press Shift-Tab once to switch to OBI 2 view.
2. Press Shift-keypad 4 to look out the left window, and then press Shift-keypad 8 to return to the forward view.

Boredom

"Lear Sierra Lima, make a right turn to a new heading of two eight zero."

"Roger, Sierra Lima, right to two eight zero."

Return to the forward view and turn off the Autopilot. Make a right turn to a new heading of 280°. Once we reach the new heading, change the Autopilot's heading hold to 280 and turn the Autopilot back on. Our new heading will have us pointed just slightly to the right of the moon. From now until the time we make contact

Aeronautically Speaking:
Ever wonder what could be stressful to a corporate or airline pilot? How about hours of sheer boredom while the instruments and computers fly the aircraft, and then an intense 15 minutes of a white-knuckle instrument landing system (ILS) approach through a thunderstorm or dense fog?

Wing Tip:
If you don't want to sit for such a long period during the flight, you can change the simulation speed of Flight Simulator by selecting Simulation Speed from the Sim menu or by clicking with the mouse on the Rate of Simulation display. This is displayed under the time on the left side of the Learjet's instrument panel. Clicking to the right of the indicated rate increases its value, and clicking to the left decreases its value. If you change the simulation rate to 4X, 1 hour of flight time will take 15 minutes. But be sure to change it back to 1X before you start to navigate into the Chicago area. If you think an ILS approach is difficult at regular speed, just try it at 4X speed!

with the Chicago Heights VOR, we will have a long flight—about an hour and a half. You can spend this time reviewing the map of Chicago, mentally preparing for the ILS approach, or just checking out the various views.

Turn Right to 280°

1. Press Z to turn off the Autopilot.
2. Turn right to a new heading of 280°, holding an altitude of 32,000.
3. After leveling out the turn, select the Nav/Com menu and select Autopilot.
4. Set HDG (Heading) Hold to **280** and select Connected (On) from the Autopilot Switch box.

"Lear Sierra Lima, maintain current course and contact Chicago Center."

"Roger, New York Center, good day…"

"…Chicago Center, this is Learjet November Five Foxtrot Sierra Lima, with you at flight level three two, heading two eight zero."

"Sierra Lima, Chicago Center. Maintain course as filed."

"Roger, Sierra Lima."

After another half hour of flying (at about 5:45 a.m.), you will notice that the heading has been slowly drifting to the left. Instead of going through the trouble of turning off the Autopilot, changing the heading, and then turning the Autopilot back on, you can make the heading change by adjusting the Autopilot: Select HDG (Heading) Hold and change it to 281. The Autopilot will automatically turn the aircraft to the desired heading of 280°. This drifting will continue to happen during the flight. Just remember to make an adjustment about every half hour.

After 6:00 a.m. the sky will begin to get lighter. By 6:30 a.m., the stars and the moon have disappeared and the sky is blue. The lights of the instrument panel will automatically turn off. Once it is daylight, you can use Map view to see the general location of the aircraft. Turn on Map view and then zoom out the view until you can see Lake Michigan. The red cross hairs in the center of the screen represent our position. We'll leave Map view on through the rest of the flight. Again, by this time you will need to make another change to the heading by changing the HDG (Heading) Hold setting in the Autopilot.

Adjust Heading and Turn On Map View

1. At 5:45 a.m., open the Nav/Com menu, and select Autopilot. Set HDG (Heading) Hold to **281**.

2. Press Num Lock to turn on Map view.

3. Press the – (minus) key at the top of the keyboard to zoom out the Map view until you can see Lake Michigan.

4. At 6:45 a.m., adjust the heading again (as in step 1 above) to 281°.

Chicago Heights VOR

"Lear Sierra Lima, Chicago Center, intercept Chicago Heights VOR and fly direct."

"Roger, Sierra Lima, direct to Chicago Heights VOR."

Around 6:48 a.m., we should make contact with the Chicago Heights VOR, which will be about 75 miles away. To determine our new heading, adjust the OBI 2 heading until the vertical needle is centered. The indicated heading at the top of the OBI instrument should be our new heading. Add another 4° to the OBI heading; this will be our new target heading. Turn off the Autopilot and make a right turn to the new heading. By the time we reach the new heading, the vertical needle of the OBI should be close to the center of the instrument. After reaching the new heading, set the heading and altitude hold of the Autopilot and then turn the Autopilot back on. You'll notice that the Map view has changed according to our change in the heading.

Intercept Chicago Heights VOR

1. After getting a signal from Chicago Heights VOR, press V and then press 2 to select the OBI instrument.

2. Use the + (plus) and – (minus) keys at the top of the keyboard to adjust the OBI heading until the vertical needle is centered.

3. Add 4° to the OBI-indicated heading for a new target heading.

4. Press Z to turn off the Autopilot.

5. Turn right to the new target heading.

6. Set the new target heading for the Autopilot and turn the Autopilot back on.

This is a good time to adjust the NAV 1 radio for the ILS approach. The frequency for the ILS for runway 27 Left is 111.1. Set the NAV 1 radio to the new frequency, but keep the OBI view on OBI 2. Now set the EFIS settings for the instrument approach. Turn on the EFIS Master Switch, Lock to ILS for Landing Approach, and Plot Intercepting Path. The Type should be Rectangles, the Density should be Thin, and the Range should be Long. When we get closer to the ILS approach path for runway 27 Left, the EFIS display will appear, and we'll be able to follow it to the runway.

Set NAV 1 and EFIS

1. Open the Nav/Com menu, and select Navigation Radios. Set NAV 1 Frequency to 111.1.
3. Open the Nav/Com menu again, and select EFIS/CFPD Display.
4. Select EFIS Master Switch, Lock to ILS for Landing Approach, and Plot Intercepting Path.
5. Set the Type to Rectangles, the Density to Thin, and the Range to Long.

"Lear Sierra Lima, Chicago Center. At 30 DME from Chicago Heights begin descent to flight level one five."

"Roger, Chicago Center. Sierra Lima descending to flight level one five."

Look at the engine instruments to be sure that there is still plenty of fuel left. Now check the distance to the VOR on your DME. At 30 miles, turn off the Autopilot and reduce the throttle to a fan setting of 60 percent. The nose of the aircraft will drop below the horizon as we start the descent. Use the elevator trim to adjust the aircraft for a descent rate of 2000 feet per minute.

Begin Descent

1. Press Tab to view the engine instruments and check the engine gauges.
2. Press Tab again to switch back to flight instruments.
3. At 30 miles from the VOR, use keypad 3 to reduce the throttle to a fan setting of 60 percent.
4. Slightly lower the nose for a descent.
5. Use keypad 7 and keypad 1 to set trim for a descent rate of 2000 feet per minute.

"Chicago Center, Lear Sierra Lima requesting heading change to three six zero."

"Lear Sierra Lima, Chicago Center. Heading change approved."

"Thank you, Sierra Lima is turning right to three six zero."

At 7 miles from the VOR, make a right turn to a new heading of 000°. This lines us up to intercept the Chicago–O'Hare ILS. We'll see Lake Michigan in the front view screen. You can also zoom in the Map view a little to get a better look at the surrounding area. Now we need to slow down a little. Extend the spoilers until the airspeed drops below 280 knots, and then retract the spoilers.

Turn to a Heading of 000°

1. At 7 miles from the VOR, make a right turn to a heading of 000° and continue descent.
2. Press + (plus) at the top of the keyboard a couple of times to zoom in for a closer view of Lake Michigan.
3. Press / (slash) on the keyboard or click on EXT on the instrument display to extend the spoilers.
4. When the airspeed drops below 280 knots, press / (slash) on the keyboard or click on RET on the instrument display to retract the spoilers.

ILS Approach

"Lear Sierra Lima, Chicago Center. Continue descent into the O'Hare TCA. Contact approach control."

"Roger, Chicago Center. Lear Sierra Lima, switching to approach control…"

"…Approach control, this is Lear November Five Foxtrot Sierra Lima, heading three six zero, descending for ILS intercept."

"Sierra Lima, approach has you on radar. Continue descent and intercept ILS approach for runway two seven left. ILS frequency is one one one point one."

"Roger, approach. Sierra Lima on intercept. One one one point one."

Change the NAV 2 radio to 113.9, the frequency of the Chicago–O'Hare VOR. We won't use it for navigating our heading, but we'll use it to keep track of our distance from the airport. As we get closer to the airport, the EFIS system will turn on and we'll be able to see the red rectangles on the front view screen. Most likely we are above and a little to the right of the EFIS path. Reduce the throttle to a fan setting of 50 percent. Switch to the OBI 1 display for viewing the glide slope. Now add 8° of flaps. Compensate for the additional lift by gently lowering the nose.

Switch to O'Hare VOR

1. Open the Nav/Com menu, and select Navigation Radios.
   Set the NAV 2 radio frequency to 113.9.
2. After the EFIS system comes on, use keypad 3 to reduce the throttle to a fan setting of 50 percent.
3. Press Shift-Tab twice to switch to the OBI 1 display.
4. Press F6 to lower flaps to 8°.
5. Compensate for the additional lift by lowering the nose.

"Chicago approach, Lear Sierra Lima is turning left for final approach for runway two seven left."

"Sierra Lima, proceed on final. You are cleared to land."

"Roger, Sierra Lima is cleared for landing."

We see from the EFIS that the intercept point for the ILS is ahead. Instead of trying to line up with the EFIS before the intercept point, we'll wait and make our adjustments and turn onto the ILS approach path. Lower the gear, and then raise the nose a little to compensate for the extra drag. Before we reach the ILS approach path, we will need to start a left turn into the approach. Try to begin and end the turn so that the aircraft intercepts the EFIS ILS rectangles. For a proper ILS approach, the aircraft should be in the center of the rectangles. Lower the flaps to 20° and adjust for the additional lift. Try to use the throttle to adjust the descent rate instead of using the yoke to raise or lower the nose. Use gradual movements to adjust for the aircraft position within the approach.

Intercept the ILS

1. Press G to lower the landing gear.
2. Raise the nose a little to compensate for the drag.
3. Make a left turn, and try to end the turn within the EFIS rectangles.
4. Press F7 to lower the flaps to 20°.
5. Gradually adjust the aircraft to line up within the EFIS rectangles.

Follow the rectangles to the runway. As we pass over the end of the runway, reduce the throttle to 0 percent, level off, and let the aircraft float to the runway. As soon as the wheels touch down, apply the thrust reversers and leave them on until the airspeed drops below 50 knots. Then use the brakes to stop.

Land the Learjet

1. Upon flying over the end of the runway, reduce the throttle to 0 percent.
2. Level off and gently float to the runway.
3. Hold down keypad 3 to apply full thrust reversers.
4. When the airspeed drops below 50 knots, use keypad 9 to return the throttle to 0 percent.
5. Press . (period) to apply the brakes until the aircraft stops.

Because you were able to get the president of the company to Chicago in time for his 9:00 meeting, he has decided to give you a raise.

✈ ADVENTURE ✈
no. 3

ENGINE OUT OVER INNSBRUCK

Things are going just fine on your flight to Munich. The weather is beautiful, and the scenery is spectacular. But just when you least expect it, your engine quits. There's no chance of clearing the mountains ahead. If you do it right, you might be able to "dead-stick" the aircraft into the Innsbruck airport, which is somewhere below, and not have the embarrassment of landing in a farm field.

Flight Plan: The engine of your Cessna R182 has quit while you are flying over the mountains bordering Germany and Austria. Because your altitude is 9000 feet, you won't be able to clear the mountain range. But, fortunately, the Innsbruck airport is somewhere below in the valley. Your assignment is to make a power-off landing at the Innsbruck airport.

✈✈✈

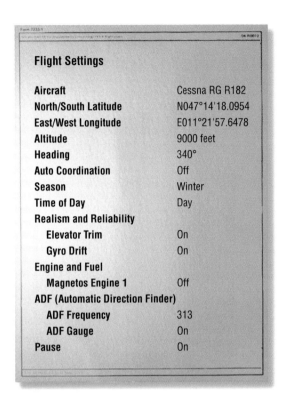

Enter the Flight Settings

1. Press P to pause Flight Simulator.

2. Open the Options menu.
   Select Aircraft.
   Select Cessna Skylane RG R182.

3. Open the World menu.
   Select Set Exact Location.
   Set North/South Lat. to **N047*14'18.0954**.
   Set East/West Lon. to **E011*21'57.6478**.
   Set Altitude to **9000**.
   Set Heading to **340**.

4. Open the World menu again.
   Select Set Time and Season.
   Set Season to Winter.
   Set Time of Day to Day.

5. Open the Sim menu, and be sure Auto Coordination is not checked.
   Select Realism and Reliability.
   Select Elevator Trim and Gyro Drift.

6. Open the Sim menu again.
   Select Engine and Fuel.
   Set Magnetos Engine 1 to Off.

7. Open the Nav/Com menu.
   Select ADF (Automatic Direction Finder).
   Set ADF Frequency to **313**.
   Select Activate ADF Gauge.

8. Open the Options menu.
   Select Save Situation.
   Enter your own description for this flight, and save this situation as **ADV3FLT**.

Oops!

Usually a few choice words would be appropriate now. As you unpause the simulator, your aircraft is instantly without power. Let the nose drop until the airspeed builds up to over 80 knots. Once the airspeed is high enough to enable you to control the aircraft, raise the nose for straight and level flight. You should be able to recover from the dive at around 8000 feet. You won't be able to fly completely level, because the engine is out, but you can establish a fairly level glide. Trim the aircraft for a constant airspeed of 80 knots, but don't take too much time doing it. Your immediate concern should be changing the heading of the aircraft. Make a right turn to a heading of 90° to avoid crashing into the Alps.

Regain Control of the Aircraft

1. Press P to unpause the simulator, and press G to raise the landing gear, letting the nose of the aircraft drop.

2. After the airspeed builds up to over 80 knots, gently pull the nose up for straight and level flight.

3. Using the keypad 7 and keypad 1 elevator trim keys, trim the aircraft for an airspeed of 80 knots.

4. Make a right standard-rate turn to a heading of 90°. (Remember to use your rudder during your turns.)

Wing Tip:
Because you cannot choose power settings in this situation, the aircraft has no forward momentum. As soon as you unpause the simulator, the nose of the aircraft pitches down toward the ground until you build up enough airspeed for gliding.

Aeronautically Speaking:
Real-world pilots don't get paid for flying. They get paid for knowing what to do in emergencies. Clear thinking, a cool head, and a knowledge of the appropriate emergency procedures are needed to keep the passengers and the aircraft safe.

Adventure 3: Engine Out over Innsbruck

Time for Questions

Now that the aircraft is under control and you are headed away from the mountains, it's time to assess the situation. To turn this emergency into a safe situation, you need to answer some questions. Your first one should be "Where am I going to land?" By looking at the Munich sectional chart in the *Flight Simulator Pilot's Handbook*, you can see that the Innsbruck airport is in the area. You did have the forethought to lock the ADF to the transmitter to the east of the airport. (The ADF gauge occupies the same location as the OBI 2 gauge on the Cessna. Pressing Shift-Tab switches back and forth between them.) The ADF gauge is pointing to the location of the transmitter, which should be slightly off to the left of the aircraft. According to the map of the area, this places the Innsbruck airport behind you.

Your next question is "What is my altitude?" Altitude becomes a crucial issue when your aircraft has no power. Your altitude should be somewhere between 6000 and 7000 feet. Now ask yourself "What is the altitude of the airport?" By knowing the altitude of the airport, you will have a good idea of the altitude of the surrounding terrain, and you will know how much altitude you have left for your glide to the airport. The airport table in the *Flight Simulator Pilot's Handbook* gives the altitude of the Innsbruck airport as 1906 feet. If your aircraft is at 6000 feet, you have a little over 4000 feet left before you reach the ground.

Because the ADF transmitter is lined up with the approach path to the Innsbruck airport, you can use it to line up with the runway. Take another look at the ADF gauge on the instrument panel. Notice that you are flying to the right of the ADF transmitter. This will give you room for a left turn over the top of the transmitter when you are heading back toward the airport. Before you make your turn, recalibrate your directional gyro to make sure that it is displaying the correct heading. When the ADF gauge shows that the transmitter is 310° from the aircraft, make a left standard-rate turn to a new heading of 260°.

Prepare for the Final Approach

1. Ask yourself the following questions:
   Where am I going to land? (Innsbruck airport)
   Where is the airport? (Somewhere behind you)
   What is my altitude? (Should be between 6000 and 7000 feet)
   What is the altitude of the airport? (1906 feet)

2. Notice the reading of the ADF gauge.
3. Press D to recalibrate the directional gyro.
4. When the ADF gauge shows the transmitter at 310°, make a left standard-rate turn to 260°.

You should now have the runway in sight. (If you don't, restart the situation.) Be sure to keep the aircraft at a constant speed of 80 knots. Depending on your distance to the airport, you might need to try to squeeze a little more distance

Adventure 3: Engine Out over Innsbruck

out of your glide. You can trim the aircraft for an airspeed of 70 knots, which will help a little. When the aircraft reaches an altitude of 3000 feet, add 10° of flaps to slow the aircraft down to about 65 knots. Trim the aircraft to keep your airspeed constant, but keep a close eye on the indicator to be sure you don't stall. (It is better to land short of the runway with the wheels on the ground than to stall and crash.)

Make the Final Approach

1. Check the airspeed to be sure it is 80 knots.
2. Use keypad 7 and keypad 1 to trim the aircraft to 70 knots if you need more distance.
3. At 3000 feet, press F6 to add 10° of flaps and trim the aircraft to 65 knots.

When the altitude reaches 2500 feet, lower the landing gear. This will slow the aircraft even more and will pitch the nose up a little. Add a little down elevator to push the nose down and keep the aircraft from stalling. As you pass over the end of the runway, begin your flare by gently pulling the nose up until the aircraft is level, and let it float to the runway as you would with a standard landing.

Prepare for Landing

1. Press G at 2500 feet to lower the landing gear.
2. After you pass over the end of the runway, flare for the landing.

Whew!

Congratulations! You made it to the runway. You kept a cool head, asked yourself a few questions, and made the aircraft do what you wanted (within its limitations, of course).

For some variations on this adventure, try it during the evening and then again at night, when the darkness plays tricks with your depth perception. Or create a cloud layer between 4000 and 7000 feet and fly through it to land on the runway. And remember, throughout any flight emergency (real or simulated), keep a cool head!

Adventure 3: Engine Out over Innsbruck

ADVENTURE no. 4

THE PILOT WHO TERRORIZED PARIS

Your persona for this adventure is Pierre the Pilot, a commercial pilot who has become tired of the rules and regulations of flying. You long for the good old days when pilots could fly anywhere they wanted and could thrill the spectators on the ground with their stunts and trick flying over the towns and farms of France. After flying a planeload of passengers into Charles de Gaulle Airport, and listening to ATC telling you what to do and where to go, you've decided that you just can't take it any more. You want to once again feel the freedom of flying without worrying about the rules and regulations, to thrill (or terrorize) the tourists and residents of Paris. Come to think of it, you remember that there is a beautifully restored Sopwith Camel parked on the ramp at Orly Airport, south of Paris…

Flight Plan: As Pierre the Pilot, you'll take off from Orly Airport in a Sopwith Camel. You'll fly under the Eiffel Tower, to the amazement of thousands of spectators, and then you'll fly between the buildings of the city. Finally, you'll land on the street and taxi under the famous Arc de Triomphe. After accomplishing all of these daredevil but illegal feats, you'll allow the authorities to arrest you.

Aeronautically Speaking: Current rules and regulations in aviation have been established for the safety and protection of both pilots and people on the ground. In the days of the Sopwith Camel, flying was a much simpler task. There were few aircraft in the air (except during wartime air battles), and none could fly faster than 160 knots, except in a dive. Today thousands of aircraft are in the air at any given moment. And with standard commercial aircraft speeds in excess of 350 knots, it is difficult to see an approaching aircraft until it is too late. Therefore, ignoring the defined rules and regulations can lead to deadly consequences. This is why Flight Simulator is so much fun. Not only can you fly any way you want, but you can always get up and walk away from a crash.

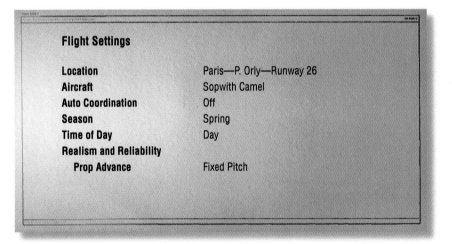

Flight Settings

Location	Paris—P. Orly—Runway 26
Aircraft	Sopwith Camel
Auto Coordination	Off
Season	Spring
Time of Day	Day
Realism and Reliability	
Prop Advance	Fixed Pitch

Enter the Flight Settings

1. Open the Options menu.
   Select Aircraft.
   Select Sopwith Camel.

2. Open the World menu.
   Select Airports.
   Set Choose A Scenery Area to FRANCE-Paris, and set Choose
      The Airport You Want to Fly From to Paris-P. Orly-Runway 26.

3. Open the World menu.
   Select Set Time and Season.
   Set Season to Spring.
   Set Time of Day to Day.

(continues on next page)

4. Open the Sim menu, and be sure Auto Coordination is not checked.
   Select Realism and Reliability.
   Set Prop Advance to Fixed Pitch.

5. Open the Options menu.
   Select Save Situation.
   Enter **ADV4FLT** in the Situation Title box. Add your own description in the Description box if you want.

Bon Voyage!

You've convinced the owner of the Sopwith Camel that you're a competent pilot and you just want to do some touch-and-gos. Apply full throttle and keep the current heading while rolling down the runway. The tail of the aircraft will come up at about 40 knots, giving you a better view of the runway. At 80 knots, gently pull back on the yoke and lift off the runway. Keep your airspeed at about 80 knots and climb to 1000 feet (the altimeter needle will be halfway between the 0 and the 2). During the climb, look out the back for a good view of the airport runways. Return to the forward view, and then level off at 1000 feet with the same takeoff heading of 258°.

Take Off

1. Press . (period) to release the parking brakes.
2. Hold down keypad 9 to apply full throttle.
3. At 80 knots, gently pull back on the yoke and lift off the runway, climbing at 80 knots.
4. Press Shift-keypad 2 to look out the back of the aircraft at the runways.
5. Press Shift-keypad 8 to return to the forward view.
6. Level off at 1000 feet with a heading of 258°.

So far, so good! Except that by now, the owner of the aircraft has noticed that you are not staying in the airport traffic pattern for touch-and-gos. But you've got more important things to do. Look out the right side of the aircraft. You should be able to see the Eiffel Tower. Return to the forward view and make a right turn to a new heading of 353°, which points the aircraft just to the right of the Eiffel Tower.

By now, the owner of the Sopwith is wondering what's going on. When you see the Eiffel Tower in front of you through your cockpit window, begin your descent. Reduce the throttle until the RPM is around 1800. Drop the nose of the aircraft slightly and maintain an airspeed of between 120 and 130 knots during the descent. Keep an eye on the Eiffel Tower. To make a safe pass under it, you will need to line up the aircraft with the arch at its base. When the aircraft is just slightly above the ground (the altimeter needle is in the middle of the 0), level off and bring the throttle back up to 2200 RPM.

Wing Tip:
Remember that in the Sopwith Camel the indicated heading is given by a magnetic compass instead of a directional gyro. In a turn, the magnetic compass lags behind the movement of the aircraft. Therefore, it is difficult to fly precision headings with the Sopwith Camel. Try to anticipate the delay in the magnetic compass and then make minor corrections to your heading once the compass becomes adjusted to your current heading.

Turn and Descend

1. Turn to a new heading of around 353°, pointing to the right of the Eiffel Tower.
2. Reduce the throttle to around 1800 RPM.
3. Lower the nose of the aircraft slightly and keep the airspeed between 120 and 130 knots.
4. When the altimeter needle is in the middle of the 0, level off the aircraft and increase the throttle to 2200 RPM.

Now the owner of the Sopwith has realized you're not coming back, and he is on his way to a telephone to report you to the French aviation authorities. Take another look at the Eiffel Tower and line up with the arch at its base. You may need to descend a little more in order to clear the first deck of the tower. The trick here is to keep clear of the support structure and yet not hit the ground. After you pass under the tower, take a look out the back.

Fly Under the Eiffel Tower

1. Line up the aircraft with the arch at the base of the Eiffel Tower.
2. Descend as needed in order to clear the first deck of the tower.
3. Press Shift-keypad 2 after passing under the tower to look out the back of the aircraft.
4. Press Shift-keypad 8 to return to the forward view.

That gave the tourists a big thrill! Not only has the owner of the aircraft called the authorities, but police at the Eiffel Tower have called in to report that a lunatic in a bright green biplane has flown under the tower and is now headed for the center of the city.

Vive la France

Now you see the buildings in the center of the city up ahead. Aim the aircraft to the right and stay just above the ground. The altimeter needle should be in the middle of the 0. After you cross over the river Seine, make a left turn and try to line up your aircraft with one of the city streets. Stay over the street as you fly between the buildings. Take a quick look out the left side of the aircraft for a view of the buildings as you fly by. After flying past the last building, take a look out the back for a nice view of the city.

Fly Through the City

1. Keeping the aircraft level and slightly above the ground, turn the aircraft to face to the right of the city.
2. After flying over the river Seine, prepare to line up on one of the city streets.
3. Pick a street and make a left turn to line up on the street.
4. Fly over the street and between the buildings of Paris.
5. Press Shift-keypad 4 to take a quick look left, and then press Shift-keypad 8 to return to the forward view.
6. Press Shift-keypad 2 for a rear view after passing the last building, and then press Shift-keypad 8 to return to the forward view.

The occupants of the buildings you buzzed are staring in disbelief. The French authorities are trying to figure out how to apprehend you. Now it's time for your culminating feat of low-level flying.

After returning to the forward view, apply full throttle. Climb back up to 1000 feet and then make a left turn to a heading of 100°. When you reach this heading and have leveled off at 1000 feet, reduce the throttle to full off. As you begin your descent, notice the road directly in front of you—the one that crosses the river. Now imagine that that stretch of road is a runway. Plan an approach and landing on the road. You may need to make a few turns to line up with it.

Line Up for the Final Approach

1. Hold down keypad 9 to apply full throttle and climb to an altitude of 1000 feet.
2. Make a left turn to a heading of 100°, and then use keypad 3 to reduce the throttle to full off.
3. Start a descent and approach for a landing on the road straight ahead that crosses the river.

Adventure 4: The Pilot Who Terrorized Paris

Wing Tip:
The Sopwith Camel can be a very difficult aircraft to land. Remember to use gentle movements with the controls, and try to keep the tail up until the airspeed drops below 40 knots. Don't try to flare the Sopwith as you would the Cessna or the Learjet, but try to keep the aircraft more level with the ground as you touch down.

As you line up with the road for a landing, you will notice that it makes a slight turn to the right. If you touch down before the turn, taxi around the turn and head down the center of the road. Otherwise, make a turn to line up with the road after it turns and plan to touch down beyond that point. After the landing, continue to taxi down the center of the road. Ahead you will see the famous Arc de Triomphe. Taxi beneath the arch and stop the Sopwith Camel just before you reach the next monument.

Land the Sopwith Camel

1. Line the aircraft up with the road for your landing.
2. Land the Sopwith Camel on the road. (Beware—you might bounce a bit!)
3. After landing, continue to taxi through the Arc de Triomphe.
4. Stop the aircraft short of the next monument.

Congratulations! *Incroyable!* Climb out of the cockpit of the aircraft and wave to the cheering crowd of Parisians and tourists until the police come and take you away. This moment will go down in history as the day Pierre the Pilot terrorized the city of Paris!

ADVENTURE no. 5

LOST AND FOUND

Okay, it's time to put up the books and find out how good you really are! Close the *Flight Simulator Pilot's Handbook* and put your maps away. The only book you'll need for this flight is the one you're currently reading. Take a look at the following flight plan and you'll see why:

Flight Plan: As your instructor, I have decided to give you the ultimate navigation test. Your flight will begin in the Learjet, at an undisclosed location. With an overcast sky, at night, with a single VOR frequency dialed into the navigation radio, you must find your way to the nearest airport and land. During this flight, you are not permitted to look at the *Flight Simulator Pilot's Handbook*. If you just can't seem to find your way, you can look at the clues on page 202.

✢✢✢

Adventure 5: Lost and Found

Flight Settings

Aircraft	Learjet 35A
North/South Latitude	N031°55'14.8150
East/West Longitude	W116°22'59.7562
Altitude	30,000 feet
Heading	032°
Auto Coordination	On
Season	Spring
Time of Day	Night
Weather	Clouds
Coverage	Overcast
Base	10,000 feet
Tops	15,000 feet
Pause	On
NAV Radio 1	117.8

Enter the Flight Settings

1. Press P to pause Flight Simulator.

2. Open the Options menu.
   Select Aircraft.
   Select Learjet 35A.

3. Open the World menu.
   Select Set Exact Location.
   Set North/South Lat. to **N031*55'14.8150**.
   Set East/West Lon. to **W116*22'59.7562**.
   Set Altitude to **30000**.
   Set Heading to **32**.

4. Open the World menu again.
   Select Set Time and Season.
   Set Season to Spring.
   Set Time of Day to Night.

5. Open the World menu again.
   Select Weather.
   Select Edit Clouds.
   Set Type to Overcast.
   Set Base to **10000**.
   Set Tops to **15000**.

6. Open the Sim menu and be sure Auto Coordination is checked.

7. Open the Nav/Com menu.
   Select Navigation Radios.
   Set NAV 1 Frequency to 117.8.

8. Open the Options menu.
   Select Save Situation.
   Enter **ADV5FLT** in the Situation Title box,
      and enter a description of your choice in the Description box.

I'm Lost!

Because you have no speed when this situation begins, the aircraft has no forward momentum. As soon as you unpause the simulator, the nose of the aircraft pitches down toward the ground until you build up enough airspeed for flight.

Let the nose drop until the airspeed builds up to over 130 knots. Once the airspeed is high enough to control the aircraft, raise the nose for straight and level flight. You should be able to recover from the dive at around 26,000 feet.

Find the Airport

1. Use keypad 9 to adjust the throttle to 80 percent.
2. Press P to unpause Flight Simulator, and press G to raise the landing gear.
3. Let the nose drop until the airspeed builds to 130 knots, and then raise the nose for straight and level flight.
4. Use your navigational knowledge to find the airport!

Spot view of the Learjet during a turn, with the moon in the background.

I'll give you one small clue to get you started: Your NAV 1 radio is tuned to a VOR transmitter. This means that your OBI 1, when properly adjusted, will give you an indication of where the VOR is located in relation to your current position. If you can find the VOR, you should be able to find a place to land.

Additional Clues

1. Adjust OBI 1 until the needle is centered. Turn to the heading displayed at the top of the OBI instrument. You will be flying toward the VOR transmitter.
2. When the DME indicates that you are 20 miles away from the VOR transmitter, reduce the throttle to 30 percent and descend to an altitude of 8000 feet. Once you are below the clouds, locate the airport and land.

ADVENTURE no. 6

LEISURE FLIGHT

One of the nice things about flying is that on a nice, clear, sunny day, you can get a few friends together and go look at the scenery from above. The San Francisco Bay area is a great place for such a flight. The weather is almost always nice (except when it's foggy), the view is beautiful, and there are lots of airports within an hour of flying.

Flight Plan: We'll fly the Cessna out of San Jose International and head out over the city of San Francisco toward the Sausalito VOR. After flying over the Golden Gate Bridge, we'll change our heading and fly to the Sacramento VOR. Then we'll head straight in for a landing at Sacramento Metropolitan Airport.

✈✈✈

Flight Settings	
Aircraft	Cessna Skylane RG R182
North/South Latitude	N037°22'13.9911
East/West Longitude	W121°56'18.5827
Altitude	59 feet
Heading	123.46°
Auto Coordination	On
Season	Spring
Time of Day	8:00 a.m.
NAV Radio 1	116.2
OBI 1 Instrument	300°
NAV Radio 2	115.2

Enter the Flight Settings

1. Open the Options menu.
   Select Aircraft.
   Select Cessna Skylane RG R182.

2. Open the World menu.
   Select Set Exact Location.
   Set North/South Lat. to **N037*22'13.9911**.
   Set East/West Lon. to **W121*56'18.5827**.
   Set Altitude to **59**.
   Set Heading to **123.46**.

3. Open the World menu again.
   Select Set Time and Season.
   Set Season to Spring.
   Set Exact Time to 8:00 a.m.

4. Open the Sim menu, and be sure Auto Coordination is checked.

5. Open the Nav/Com menu.
   Select Navigation Radios.
   Set NAV 1 Frequency to **116.2** with an OBS Heading of **300**.
   Set NAV 2 Frequency to **115.2**.

6. Open the Options menu.
   Select Save Situation.
   Save this situation as **ADV6FLT**.
   Enter a description of your choice in the Description box.

Flight Planning and Takeoff

Even though this is a leisure flight, it is still a good idea to do a little flight planning. On the sectional chart in the *Flight Simulator Pilot's Handbook* for the San Francisco area, find the San Jose International Airport. We'll be taking off from there and then heading north to the Sausalito VOR, located north of the Golden Gate Bridge. From there, we'll change our heading and fly toward the Sacramento VOR. After passing over the top of the Sacramento VOR, we'll turn left and head straight in to Sacramento Metropolitan Airport.

Again, I'll take care of the radio communications for us.

"San Jose Tower, this is Cessna November Two Oscar Oscar One Zulu, in position on runway one two."

"Roger, One Zulu, you are cleared for takeoff. After departure, make a right turn to heading three zero zero."

"One Zulu, taking off and turning right to three zero zero."

Release the parking brakes, apply full throttle, and keep the plane centered down the runway. Rotate at 80 knots, raise the gear, and then lower the nose a little to help the airspeed build up. Trim the aircraft for a climb speed of 80 knots. You should see the horizon just above the top of the instrument panel. Now make a right standard-rate turn to a heading of 300°. Remember to start coming out of your turn at 290° so that when you've finished the turn the heading will be 300°.

Adventure 6: Leisure Flight

Aeronautically Speaking:
Each VOR has a three-letter identification code. The Sausalito VOR is referred to as the SAU VOR. The San Francisco VOR is referred to as the SFO VOR.

Aeronautically Speaking:
Flight Following is for VFR flights and allows the pilot to stay in close contact with the ground controllers. Radio assistance is available for traffic sequencing and checking location by radar, and the controllers may contact the pilot and request a change in flight direction or altitude. This gives the pilot a little extra comfort in knowing that someone is watching and keeping track of the aircraft even though it is not an IFR flight.

Take Off

1. Press . (period) to disengage the parking brakes, and then use keypad 9 to apply full throttle, gently pulling back on the yoke at 80 knots.
2. Press G to raise the gear after takeoff, and use keypad 7 and keypad 1 to trim for a climb speed of 80 knots.
3. While still climbing, make a standard-rate right turn to a heading of 300°.

"Cessna One Zulu, continue climbing to six thousand. Proceed VFR to Sausalito VOR and contact San Francisco Center for Flight Following."

"Roger, San Jose. One Zulu climbing to six thousand, contacting San Francisco Center. Thanks…"

"…San Francisco Center, this is Cessna November Two Oscar Oscar One Zulu, climbing to six thousand, heading three zero zero. Requesting Flight Following."

"Cessna One Zulu, continue climb as stated. Proceed as filed to SAU VOR. Flight Following approved."

"Roger, Center. One Zulu on Flight Following."

At the end of your turn, the needle of OBI 1 should be close to the center. Don't worry if it is off a bit. Because our flight is mostly visual, we won't be relying completely on the VORs. Continue the climb to an altitude of 6000 feet. During the climb, you can switch to Spot view and pan around for a nice view of the San Jose airport and the region at the south end of San Francisco Bay. After you've viewed the scenery, return to Cockpit view. Our current course has the aircraft pointed toward the Bay. After we reach our target altitude of 6000 feet, level off and reduce the power to around 75 percent. Trim the aircraft for level flight. After you level off, the airspeed will increase to about 135 knots and, depending on how long the climb to altitude takes, DME 1 should show that you are about 33 miles from the Sausalito VOR.

Climb Out

1. Continue climbing to an altitude of 6000 feet.
2. Press S twice to switch to Spot view during the climb, and use the Shift and keypad keys to pan to the various views.
3. Press S once to switch back to Cockpit view.
4. Use keypad 3 at 6000 feet to reduce the throttle to 75 percent and trim the aircraft for level flight.

The Scenic Route

By this time you should be able to see the San Mateo bridge, spanning the southern part of San Francisco Bay. Just to the north of the bridge, on the east side of the Bay, is the Hayward airport. San Carlos airport is visible off to your left. Soon you'll be able to see San Francisco International Airport on the left. To get a nice view of the city and the Golden Gate Bridge, change your heading to take us over the city.

"San Francisco Center, Cessna One Zulu is requesting course change to two seven zero for a more scenic route."

"One Zulu, course change approved. Be aware that we may ask you to change course if traffic becomes too heavy."

"Roger, Center. We appreciate the notice. Turning to two seven zero."

Make a left turn to a new heading of 270°. This will put us on a course directly over San Francisco International Airport. The needle on OBI 1 will drift off to the right, but we'll ignore it for now. You should soon hear the outer marker beacon for the approach to SFO. Looking off to the right, you should be able to see Oakland International. At about 17 miles (according to DME 1) from the SAU VOR, you'll hear the inner marker beacon.

Turn Toward San Francisco International Airport

1. Use Shift-keypad 4 or Shift-keypad 7 to look for San Francisco International Airport.
2. Make a left turn to a heading of 270° to fly directly over San Francisco International Airport.

"San Francisco Center, Cessna One Zulu requesting heading change to three two zero."

"One Zulu, heading change is approved. After reaching the Golden Gate, proceed direct to SAU VOR."

"One Zulu, turning to three two zero. Will proceed direct to SAU VOR after passing Golden Gate."

As we fly over San Francisco International, make a right turn to a new heading of 320°. This puts us on a course over the city of San Francisco. Again, we'll have a good view of Oakland International out the right window. We can also see the Oakland Bay Bridge. Looking forward, we'll be able to see the Golden Gate Bridge and the buildings of San Francisco. Continue on your current course, looking through the front view screen until you can see the Golden Gate Bridge pass below you. Adjust OBI 1 until the needle is centered. Note the heading at the top of OBI 1, and make a right turn to that heading. This will take us to the Sausalito VOR.

Turn Right Toward San Francisco

1. After passing over San Francisco International, turn right to a heading of 320°.
2. Press Shift-keypad 6 for a view of Oakland International, and then press Shift-keypad 8 to return to the forward view.

3. After passing over the Golden Gate Bridge, press V and then 1 to select OBI 1.
4. Use the – (minus) and + (plus) keys at the top of the keyboard to adjust the OBI heading until the needle is centered.
5. Turn to the heading indicated at the top of OBI 1.

On to Sacramento

"San Francisco Center, Cessna One Zulu is requesting course change direct to Sacramento VOR."

"One Zulu, course change approved."

At about 2 miles from the Sausalito VOR, tune OBI 2 until the needle is centered. Note the heading displayed at the top of the OBI 2 instrument, and then make a right standard-rate turn to the indicated heading. After coming out of the turn, we should be about 55 miles from the Sacramento VOR. This is another good time to enjoy the scenery.

Turn Right to the Sacramento VOR

1. Press V and then 2 to select OBI 2, and use the – (minus) and + (plus) keys at the top of the keyboard to adjust the OBI heading until the needle is centered.
2. Turn right to the heading indicated at the top of OBI 2.

Adventure 6: Leisure Flight

The highway up ahead, US 80, leads to Sacramento, but it takes a more northerly route than we want to take. If you were having difficulty using the VORs, you could rely on "the poor man's IFR" (I Follow Roads). But because you have a good idea of how VORs work, you should be able to navigate to Sacramento without any problems.

Notice that as you head farther north the directional gyro drifts down. If you started toward the Sacramento VOR at a heading of 38°, you'll see the instrument drift to 34° by the time you reach the VOR. As we get closer to the VOR, you'll need to readjust your heading.

This is a good time to take another look at the sectional of the San Francisco area in the *Flight Simulator Pilot's Handbook* and note the orientation of the runway at Sacramento. You need to know what heading to take from the Sacramento VOR in order to locate the airport and line up with the runway. Look at Sacramento Metropolitan Airport on the map, and then look at the Sacramento VOR. Can you tell what the heading should be from the VOR in order to reach the airport? If you guessed 340°, you're close. That heading would take you to the airport, but at a slight angle. A better heading would be 330°, which would let you locate the airport visually and have enough distance to line up with the runway. So, after you pass over the Sacramento VOR, turn to a new heading of 330°.

Review the Map

1. Review the sectional map of the San Francisco area and note the location of Sacramento Metropolitan Airport and the heading of the runway.
2. Adjust your heading to compensate for instrument drift.

When you are 29 miles from the Sacramento VOR (according to DME 2), you can see some farm fields up ahead. On this leg of the flight, there isn't much spectacular scenery, but you could use the extra time to review the map again, check the engine instruments, or just look out the windows.

"San Francisco Center, Cessna One Zulu is requesting approach to Sacramento Metropolitan from the Sacramento VOR."

"One Zulu, Center. Proceed to Sacramento VOR, then turn left heading three three zero. Contact Sacramento Metropolitan Approach. Good day."

"One Zulu, new heading will be three three zero. Contacting approach control. Thanks."

At 6 miles from the Sacramento VOR, you'll be able to see the Sacramento Executive Airport ahead. As you pass over the VOR, the approach lights for that airport come into view. This would be a good alternative airport in case of an emergency. As you approach the VOR, the OBI 2 needle will swing off to one side, and the "To" indicator will change to "Off" and then to "From." This means that you have flown over the top of the VOR. Now make your left turn to a new heading of 330°.

Approach the Airport

1. After passing over the Sacramento VOR, make a left turn to a heading of 330°.
2. After completing your turn, feel free to switch to an external view—and take a flight photograph.

The Approach

"Sacramento Metro Approach, this is Cessna November Two Oscar Oscar One Zulu, inbound from the Sacramento VOR on heading three three zero."

"One Zulu, Approach. Descend to one five zero zero and continue on current heading. Report when you have the airport in sight."

"Roger, One Zulu descending to one five zero zero."

We are a bit high for our approach, so drop the throttle to 25 percent. Let the nose of the plane drop for the descent. As we get closer to the airport, you'll be able to see it off in the distance as a thin gray line.

Adventure 6: Leisure Flight

Aeronautically Speaking: VASI (visual approach slope indicator) lights are used for visual approaches to the runway. They are usually located off to the left side of the runway but are sometimes on both sides of the runway. Two lights are visible, one above the other. If both lights are white, the aircraft's approach is too high. If both lights are red, the aircraft is too low. If one light is white and the other is red, the aircraft is at the correct approach height.

Reduce the Throttle

1. Press keypad 3 to reduce the throttle to 25 percent.
2. Visually locate the airport through the front view screen.

"Approach Control, One Zulu has the airport in sight."

"One Zulu, you are cleared for landing on runway three four left. Three four right is currently closed for maintenance."

"Roger, Approach. One Zulu cleared for straight in to three four left."

Continue on your current heading until it looks as though you can make a right turn to line up with the runway on the left. When the airspeed is within the white arc of the airspeed indicator, apply 10° of flaps and lower the gear. As you get closer, you'll be able to see the approach lights and the VASI lights off to the left side of the runway. Turn right to line up with the runway. Apply another notch of flaps. Use the throttle to adjust the descent rate of the aircraft and to stay at the correct approach height. Try to keep the airspeed around 80 knots during the approach. As you pass over the end of the runway, cut the throttle. At just a few feet over the runway, gently pull back on the yoke and let the aircraft float down to the runway. Apply the brakes after touchdown.

Land the Aircraft

1. Press F6 for 10° of flaps, press G to lower the gear, and line up with the runway.
2. Press F7 for 30° of flaps, and descend to the runway at 80 knots.
3. Cut the throttle over the end of the runway. Gently pull back on the yoke and float to the runway.
4. Press . (period) to apply the brakes when the airspeed is at 40 knots.

Congratulations! Another successful flight.

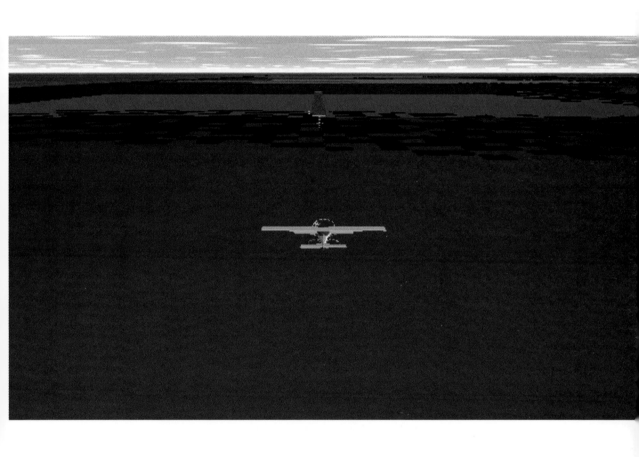

Appendix
Recommended System Requirements

Flight Simulator 5 will operate properly only if your computer system meets specific hardware and software requirements. Naturally, a fast computer with a lot of memory will enhance this program greatly, but Flight Simulator 5 is very flexible with its setup options and preferences. If you have a slow machine, you can trade display splendor for speed, or vice-versa. Use the following recommendations as guidelines for system selection and setup, and refer to your Flight Simulator 5 manual to set up the program to work best with your system.

Hardware

The following lists describe the recommended hardware requirements for Flight Simulator 5.

Basic System Hardware

- 80386, 80486, or Pentium IBM PC or compatible computer.
- Hard drive with 14 MB of free disk space.
- 530 KB of free conventional memory and 1.5 MB of extended or expanded memory. (The minimum memory requirements for Flight Simulator 5 are 530 KB of free conventional memory and 1 MB of extended or expanded memory.)
- 3.5-inch high-density floppy disk drive.
- Video Graphics Array (VGA) or Super Video Graphics Array (SVGA) display. (The minimum video requirement for Flight Simulator 5 is an Enhanced Graphics Adapter, or EGA, with 256 KB of video memory.)

Additional Hardware Options

- Sound card.
- Microsoft mouse or other compatible pointing device.
- One or two joysticks or a yoke control connected to a game port.

Software

The following describes the software needs of Flight Simulator 5 and, for users of MS-DOS 5.0 and later, recommended memory provisions.

Basic Software Requirements

- MS-DOS or PC-DOS version 3.2 or later.
- A high memory manager, such as MS-DOS HIMEM.SYS, to make extended memory available.
- An expanded memory manager, such as MS-DOS EMM386.EXE, to make expanded memory available.

Configuring Your Memory Needs with MS-DOS

If you are using MS-DOS 5.0 or later, you can gain the best performance with Flight Simulator 5 by providing 2 MB of expanded memory. You can do this by modifying your CONFIG.SYS file so that it contains the following lines:

    device=c:\dos\himem.sys

    device=c:\dos\emm386.exe 2048 ram

For more information on maximizing your system's memory, refer to your MS-DOS manual. If you are using a different memory manager, refer to the manual for that memory manager.

Index

A

ADF (automatic direction finder), 188, 189
ailerons
  control yoke and, 14
  defined, 7
  noncoordinated rudder and, 101
Aircraft Ground Traffic option, 156
Aircraft Texture option, 144
Airframe Damage from Stress option, 101
Airport Service Traffic option, 156
air pressure, 89–90
airspeed, 31
  leveling off and, 21
  temperature and, 88
airspeed indicator, 31
air traffic control (ATC)
  departure control, 106
  flight plan and, 3, 108
Air Traffic option, 156
air traffic pattern. *See* pattern flying
altimeter
  adjusting, 89–90
  described, 32
  drifting of, 89
  importance of keeping an eye on, 117
  introduced, 20
altitude
  during introductory flight, 20
  in power-off situation, 188
  rapid loss of (slip), 101
altostratus clouds, 75
angle of attack
  defined, 51
  in slow flight, 52

approach
  to Meigs Field (Chicago), 24, 26, 58–63
  midflight planning of, 129–33
  to Oakland International Airport, 133–35
  in power-off landing at Innsbruck, 188–90
  to Sacramento Metropolitan Airport, 211–12
Approach Lighting option, 150
artificial horizon, 54
artificial horizon instrument. *See* attitude indicator
ATC. *See* air traffic control
ATIS (Automatic Terminal Information Service)
  setting the Communication Radio frequency to, 10
  weather briefing, 106
attitude, 6
attitude indicator (artificial horizon instrument), 54, 140
Auto Coordination
  defined, 14
  turning off, 101
automatic direction finder (ADF), 188, 189
Autopilot, 126–27, 141, 176–79

B

banks, 7, 22–23
  Auto Coordination and, 14
boredom, 177
brakes, xx

Index   217

buzzing
  the Eiffel Tower, 44–45
  the pilot, in radio-controlled flight, 170–71
  Seattle (Washington), 154–55

C

carburetor heat
  control, keystrokes for, *xx*
  during landing approaches, 62
  testing, 15
  turning off, 9
Catalina Island, flight from John Wayne Airport to. *See* John Wayne Airport (Santa Ana, California)–Catalina Island flight
CFPD (command flight path display), 131–32
Chicago. *See* Meigs Field
cirrus clouds, 75
climbing turns, 108
climbs
  in a Learjet, 140
  pitch and, 21
  power (with throttle), 37–38
  rate of, 33
  vertical speed indicator and, 33
clouds, 75–83
  color of, 76
  settings for, 81
  types of, 75
COM (communication radio), setting frequency to ATIS, 10
command flight path display (CFPD), 131–32
compass, magnetic, 195

CompuServe, *xxiii*
controls. *See also specific controls*
  introduced, *xviii–xx*
  sensitivity of, 38
  testing the movement of, 13
control surfaces
  checking, 13
  described, 6–7
conventions, *xviii*
corporate pilot adventure (New York–Chicago), 172–84
  adjusting heading and turning on Map view, 179
  autopilot, 176–79
  beginning descent, 180–81
  departure, 174–75
  flight plan, 173
  flight settings, 173–74
  ILS approach, 181–84
  intercepting Chicago Heights VOR, 179–80
  landing, 184
  right turn, 178
  setting NAV 1 and EFIS, 180
critical angle of attack, 51
cumulonimbus clouds, 75
cumulus clouds, 75

D

departure. *See* takeoff; *specific flights*
departure control, ATC, 106
descents. *See also* landings
  through clouds, 116
  vertical speed indicator and, 33
directional gyro, 32

display preferences, 143–47
  settings for optimum performance, 144–46
  settings for the best graphics, 146–47
display refresh rate, 142, 145, 147
DME (distance measuring equipment), 109
dynamic scenery, 155–58

E

Earth Pattern option, 150
EFIS (electronic flight instrument system), 131–33
EGA graphics mode, 141
Eiffel Tower. *See* Paris (France)
electronic flight instrument system (EFIS), 131–33
elevator
  described, 6, 7
  yoke and, 13–14
elevator trim, *xx*, 20, 101
emergencies, 187. *See also* Innsbruck, power-off landing at
engine, starting, 10
Engine Stops When Out of Fuel option, 101
external views, flight photographs and, 43

F

Federal Aviation Administration (FAA), 29
finding the nearest airport, 199–202
flaps
  applying during landing approach, 61–62

flaps, *continued*
  applying at slow speeds, 57
  control, keystrokes for, *xx*
  inspecting, 5–6
flare, 62
Flicker/Speed setting, 143
Flight Following, 206
flight instruments, *xvi*, 31–33. *See also specific instruments*
flight photographs, 43, 52, 53, 159
flight plan
  "closing," on arrival, 3
  formation flying from Meigs Field (Chicago), 157–58
  for IFR flights, 108
  illegal flight over Paris, 193
  John F. Kennedy International Airport (New York) flight, 90
  John F. Kennedy International Airport (New York)–O'Hare Airport (Chicago) flight, 173
  John Wayne Airport–Catalina Island flight, 102
  Meigs Field (Chicago) flights
    air traffic pattern flight, 64
    approach flight, 58
    cloud coverage flight, 77
    local flight, 3
    windy conditions flight (vertical takeoff), 86
  Olympia Airport (Washington) flight, 138
  Paris flight in a Cessna, 29
  power-off landing at Innsbruck, 185
  radio-controlled flight, 164–71
  Reno–Oakland flight, 90, 119

flight plan, *continued*
  Rocky Mountains flight, 46
  San Jose–Sacramento flight, 203, 205
flight settings
  described, *xxi*
  illegal flight over Paris, 193–94
  John F. Kennedy International Airport (New York) flight, 91–92
  John F. Kennedy International Airport (New York)–O'Hare Airport (Chicago) flight, 173–74
  John Wayne Airport–Catalina Island flight, 103, 115
  Meigs Field (Chicago) flights
    air traffic pattern flight, 65–66
    approach flight, 58–59
    cloud coverage flight, 77–79
    local flight, 3–4
    windy conditions (vertical takeoff) flight, 86
  Olympia Airport (Washington) flight, 139
  Paris flight in a Cessna, 29–30
  power-off landing at Innsbruck, 186–87
  radio-controlled flight, 165–66
  Reno–Oakland flight, 119–21
  Rocky Mountains flight, 46–47
  San Jose–Sacramento flight, 204–5
flight simulation clubs and organizations, *xxiii–xxiv*
formation flying, 157–58
fuel gauges, 5, 6
fuel selector, 9
Fuel Tank Selector option, 102

G

glides, power-off, 60–63, 187
glide slope, 129, 134, 182
Gradient Horizon option, 144
graphics, 138, 141–50
  display preferences, 143–47
  modes, 141–42
  multiple views, 147–50
Gyro Drift option, 101

H

heading indicator. *See* directional gyro
Horizon Only (No Scenery) option, 150
humidity, 88

I

IFR (instrument flight rules)
  defined, 100
  flight plan and, 108
  situational awareness and, 117
ILS (instrument landing system) approach
  in John F. Kennedy International Airport (New York)–O'Hare Airport (Chicago) flight, 181–84
  in Reno–Oakland flight, 129
Image Complexity settings, 150
Image Quality/Speed setting, 143
Image Smoothing option, 150
inner marker beacon, 129, 207
Innsbruck, power-off landing at, 185–91
  final approach, 188–90
  flight plan, 185
  flight settings, 186–87
  landing, 191

Innsbruck, power-off landing at, *continued*
  regaining control of the aircraft, 187
inside loop, 168–69
inspecting the airplane. *See* preflight procedures
Instant Replay command, 45
Instrument Lights option, 102
instrument panel
  overview, *xx*
  red lights on, 50
  Sopwith Camel, 194
instruments. *See also* instrument panel; *specific instruments*
  relying on, 117
introductory flight. *See* Meigs Field (Chicago), local flight
inverted fly-by, 171

J

jet streams, 84–85
John F. Kennedy International Airport (New York)
  flight from, 90–96
  flight to O'Hare Airport (Chicago) from (*see* corporate pilot adventure (New York–Chicago))
John Wayne Airport (Santa Ana, California)–Catalina Island flight, 102–17
  approach into Santa Catalina airport, 114–16
  calculating the aircraft's exact location, 111–14
  climbing turn, 108
  descending through clouds, 116

John Wayne Airport (Santa Ana, California)–Catalina Island flight, *continued*
  flight plan, 102
  flight settings, 103, 115
  flight summary, 117
  intercepting vector 21, 114
  landing, 116
  navigation instruments, 109–12
  review of Los Angeles sectional chart, 111
  takeoff, 106–7
  weather briefing, 106
joystick, rudder control with, 101

K

keyboard controls and shortcuts
  for coordinated turns, 101
  for elevator trim, 20
  overview, *xix-xx*
  for rudder control, 101
  for Spot views, 7
  to toggle between maximized and minimized windows, 11
  for various views, 19

L

landing
  at Catalina Airport, 116
  in John F. Kennedy International Airport (New York)–O'Hare Airport (Chicago) flight, 184
  at Oakland Airport, 136
  "perfect," 60
  power-off, at Innsbruck, 191
  in radio-controlled flight, 167–68

landing, *continued*
    in San Jose–Sacramento flight, 212
    stalls and, 55
landing configuration, stall in, 56–57
landing gear, *xx*, 48
Landing Lights Available option, 143
Land Me feature, 26
Learjet controlling, 93, 121, 140, 141.
    See also corporate pilot adventure
    (New York–Chicago)
leisure flight, 203–13. *See also*
    San Jose–Sacramento flight
lessons, 28
level flight (leveling off), 83. *See also*
    straight and level flight
    airspeed and, 21
    trimming the aircraft for, 20–21
    visually checking for, 21–22
liftoff, 17, 18
Lights Burn Out option, 102
loading situations, *xxii*
location, calculating, 111–14
Location Readout setting, 143

M

magnetic compass, 195
magnetos, 5, 6, 14–15
Map Display setting, 143
Map view, 149, 150, 179–80
marker beacons, 35
Meigs Field (Chicago)
    air traffic pattern flight, 64–70
        base leg, 70
        crosswind leg, 67
        downwind leg, 67–68
        final approach and landing, 70–71

Meigs Field (Chicago), air traffic
    pattern flight, *continued*
        flight settings, 65–66
        takeoff leg, 66–67
    approach to, 24, 26, 58–63
    cloud coverage flight, 77–83
        cloud coverage settings, 80
        flight settings, 77–79
        leveling off, 83
        takeoff and turning climb, 80
    formation flight, 157–58
    landing at, 58–63
    local flight, 18, 24–26
        approach, 24, 26
        cruising over Chicago, 24
        departure, 16–17
        flight plan, 3
        flight settings, 3–4
        leveling off, 20–22
        preflight procedures, 2–10
        turning and banking, 22–23
        walk-around, 4–9
    windy conditions (vertical takeoff)
        flight, 86–88
memory requirements, *xvii*
MicroWINGS, *xxiv*
midair collisions. *See also* near misses
    prevention of, 106, 108
middle marker beacon, 35, 129
Moonlight At Night option, 150
multiple views, 147–50

N

navigation instruments, 109–12. *See
    also specific instruments*
navigation lights, 8

NAV (navigation radio), 109, 110
near misses, 36
night flights, 50. *See also* corporate pilot adventure (New York–Chicago)

O

Oakland, flight from Reno to. *See* Reno (Nevada)–Oakland (California) flight
OBI (omni-bearing indicator), 109, 110
   adjusting headings on, 113, 125, 127
oil pressure gauge, 10
oil temperature gauge, 10
Olympia Airport (Washington) flight
   Autopilot, 141
   buzzing Seattle, 154–55
   flight plan, 138
   flight settings, 139
   flying to downtown Seattle, 152
   maximum scenery, 152
   minimum scenery, 153–54
   reducing speed and altitude, 152
   setting the time, 152
   takeoff, 140
outer marker beacon, 36, 129, 207

P

Paris (France)
   illegal flight, 192–98
      final approach, 197–98
      flight plan, 193
      flight settings, 193–94
      flying through the city, 196–97
      flying under the Eiffel Tower, 196

Paris (France), illegal flight, *continued*
      landing, 198
      takeoff, 194–95
      turning and descending, 196
   local flight in a Cessna, 29–41
      buzzing the Eiffel Tower, 44–45
      climbing out after takeoff, 34
      flight plan, 29
      power climb, 37–38
      sight-seeing, 38–41
      straight and level flight, 35–36
      takeoff, 34
      turns around the Eiffel Tower, 41–43
      two-minute turns, 36–37
parking brakes, 9
pattern flying, 64–71
pausing flight simulator, *xvii*, 19, 49
PCX files, 43
peripheral view, 147, 148
photographs. *See* flight photographs
pictures. *See* flight photographs
pitch, 6. *See also* level flight (leveling off)
Pitot tube, 31
porpoising, 51
position. *See* location, calculating
power-off glides, 60–63, 187
power-off landing. *See* Innsbruck, power-off landing at
power-off stalls, in landing configuration, 56–57
preflight inspection, 3
preflight procedures, 2–10
private pilot certificate, 29

R

radio-controlled flight, 164–71
　buzzing the pilot, 170–71
　flight settings, 165–66
　inside loop, 168–69
　inverted fly-by, 171
　landing, 167–68
　takeoff, 166–67
radio frequencies, changing, with the mouse, 10
rate-of-climb indicator, 33
Realism settings, 101–2
Reno (Nevada)–Oakland (California) flight, 119–36
　adjusting the OBI, 125, 127
　Autopilot, 126–28
　checking the engine instruments, 127
　final approach to Oakland, 133–35
　flight plan, 119
　flight settings, 119–21
　landing, 136
　midflight approach planning, 129–33
　pre-takeoff considerations, 122
　takeoff, 124–25
resizing views, 147–48
Rocky Mountains flight, 46–57
RPM (revolutions per minute)
　testing carburetor heat and, 15
　testing magnetos and, 14, 15
rudder
　Auto Coordination and, 14
　controlling, *xix*, 14
　described, 6–7
　noncoordinated ailerons and, 101
runup, 12–15
runup area, 12
runway, taxiing to, 16–17

S

San Jose–Sacramento (California) flight, 203–13
　climbing out, 207
　flight plan, 203, 205
　takeoff, 205–6
saving situations, *xxii*, 12
scenery
　available parameters for, 150
　dynamic, 155–58
　maximum, 151–52
　minimum, 153–54
Scenery Frequency option, 156
screen refresh rate. *See* display refresh rate
search for missing aircraft, 3
Seattle (Washington), buzzing, 154–55
sectional charts, 104, 113
See Aircraft Shadows option, 143
See Ground Scenery Shadows option, 143
See Own Aircraft From Cockpit option, 144
See Propeller option, 143
Sensitivity setting, 38
shallow turns, 54–55
Simulation Speed option, 178
situational awareness, 117
slip, 101
slow flight, 49–52
　at 30-percent throttle, 51–52
　at 50-percent throttle, 50–51
slow turns, 54–55
Smooth Transition View option, 144
solo flight, 29

Sopwith Camel. *See also* Paris
    (France), illegal flight
  inside loop with, 168–69
  instrument panel, 194
  landing, 198
spark plugs, 14
stalls
  angle of attack and, 51
  in landing configuration, 56–57
  at 30-percent power, 55–56
  temperature and, 88
stall warning buzzer, 55
Stars In the Sky option, 150
straight and level flight. *See also* level
    flight (leveling off)
  at 50-percent throttle, 50–51
  at 80-percent throttle, 49
  sight-seeing over Paris, 38–40
  smooth movements and, 38
  visually checking for, 21–22, 35–36
stratus clouds, 75
SVGA Board Maker option, 143
SVGA graphics mode, 141–42

T

takeoff
  in clear weather, 95
  with cloud coverage, 80
  in formation, 158
  illegal flight over Paris, 194–95
  John F. Kennedy International
      Airport (New York)–O'Hare
      Airport (Chicago) flight, 174–75
  John Wayne Airport–Catalina Island
      flight, 106–7
  in a Learjet, 140

takeoff, *continued*
  Meigs Field (Chicago) local flight,
      17–18
  Paris flight in a Cessna, 34
  radio-controlled flight, 166–67
  Reno–Oakland flight, 124–25
  San Jose–Sacramento flight, 205–6
  vertical, 86–88
taxiing
  to runup area, 10–12
  to runway, 16–17
taxiway, 11
temperature, 88–89
Text Presentation setting, 143
Textured Buildings option, 144
Textured Ground option, 144
Textured Sky option, 144
throttle
  climbs and, 37–38
  controlling, xix–xx, 20
  starting the engine, 10
thunderstorms, 76
touch-and-go, 70
Traffic Outside Airport option, 156
trim, elevator, xx, 20–21, 101
turn coordinator, 32–33
turns
  around a point, 41–43
  Auto Coordination and, 14
  climbing, 108
  with cloud coverage, 83
  coming out of, 37
  introduced, 22–23
  in a Learjet, 140–41
  slow, 54–55
  two-minute, 36–37

Index 225

U

uncontrolled airspace, 36
upside-down flying, 171

V

VASI (visual approach slope indicator) lights, 212
vertical speed indicator
 "chasing the needle," 35–36
 described, 33
vertical takeoff, 86–88
VFR (visual flight rules), 75
VGA graphics mode, 141, 142
video
 playing, 161
 recording, 160–61
views
 multiple, 147–50
 resizing, 147–48
 for taxiing, 11
 for walk-around, 4, 7–8
VORs (very high frequency omni-directional ranges), 109–13, 206

W

walk-around, 4–9
weather, 74–98. *See also* air pressure; clouds; temperature; wind
weather areas, 90–94
weather briefings, 106
Weather option, 88
weather information, from ATIS, 10
wind, 80, 84–88
Wire Frame option, 150

Y

Yaw, 14
Yoke, *xviii–xviiii*, 13–14

Z

Zoom, 53
Zoom indicator, 87

About the Author

Timothy L. Trimble wandered aimlessly through his life until about 1980. Then, one day, when he was meandering through an electronics store, a personal computer caught his attention. He was immediately stricken by computer fever and had found his ultimate goal in life: to write video games. He dragged his wife and kids from the comfort of family and friends in Indiana to pursue "computer gold" in the hills of California.

After spending some years developing software for Columbia Pictures and Ashton-Tate, he has now achieved his goal of becoming a freelance writer and software developer. Tim continues to write as "The Timinator" for *Computer Gaming World* and *MicroWINGS Magazine*. He has also written for *Plane & Pilot*, *Amiga World*, and *DBMS* magazine. When he is not writing or playing Flight Simulator, he can be found buzzing the skies of San Diego in a Cessna 150.

Correspondence to "The Timinator" can be sent through CompuServe to address 76306,1115.

About This Book

The manuscript for this book was prepared and submitted to Microsoft Press in electronic form. Text files were prepared using Microsoft Word 2.0 for Windows. Pages were composed by The Leonhardt Group using Quark Express 3.11, with text in Times Roman and Helvetica Condensed. Composed pages were delivered to the printer as electronic prepress files.

Microsoft Press Graphics Design Liaison, Kim Eggleston.

Cover and Book Design by The Leonhardt Group, Seattle, Washington.

Train Yourself
with *Step by Step* books from Microsoft Press

Step by Step books are the perfect self-paced training solution for the on-the-go businessperson. Whether you are a new user or you're upgrading from a previous version of the software, *Step by Step* books can teach you exactly what you need to know to get the most from your new software. The lessons are modular, example-rich, and fully integrated with the timesaving practice files on the disk. If you're too busy to attend a class or if classroom training doesn't make sense for you or your office, you can build the computer skills you need with the *Step by Step* books from Microsoft Press.

Microsoft® MS-DOS® 6 Step by Step

Catapult, Inc.
Revised for version 6.2.
304 pages, softcover with one 3.5-inch disk
$29.95 ($39.95 Canada)   ISBN 1-55615-635-9

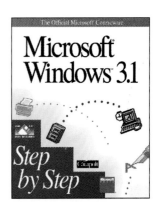

Microsoft® Windows™ 3.1 Step by Step

Catapult, Inc.
296 pages, softcover with one 3.5-inch disk
$29.95 ($39.95 Canada)   ISBN 1-55615-501-8

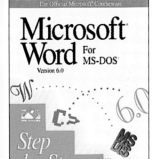

Microsoft® Word for MS-DOS® 6 Step by Step

Microsoft Corporation
272 pages, softcover with one 3.5-inch disk
$29.95 ($39.95 Canada)   ISBN 1-55615-520-4

Microsoft® Excel 5 for Windows™ Step by Step

Catapult, Inc.
368 pages, softcover with one 3.5-inch disk
$29.95 ($39.95 Canada)   ISBN 1-55615-587-5

*Microsoft*Press

Microsoft Press® books are available wherever quality books are sold and through CompuServe's Electronic Mall—GO MSP. Call 1-800-MSPRESS for direct ordering information or for placing credit card orders.*
Please refer to BBK when placing your order. Prices subject to change.

*In Canada, contact Macmillan Canada, Attn: Microsoft Press Dept., 164 Commander Blvd., Agincourt, Ontario, Canada M1S 3C7, or call (416) 293-8464, ext. 340.
Outside the U.S. and Canada, write to International Sales, Microsoft Press, One Microsoft Way, Redmond, WA 98052-6399.

Learn More About MS-DOS®

Running MS-DOS®
Covers through version 6.2

Van Wolverton

"A book even the PC mavens turn to, it is written by a human being for human beings, in a strange and wonderful tongue: English." **PC Week**

RUNNING MS-DOS, with more than 3 million readers, is the most highly acclaimed introduction and complete reference to MS-DOS available. This special tenth anniversary edition covers MS-DOS 3.3 through version 6.2. It's the sure way to gain the solid grounding in computing fundamentals that will help you better understand and work with other applications. Contains a wealth of easy-to-follow examples, instructions, and exercises.
640 pages, softcover $19.95 ($24.95 Canada)   ISBN 1-55615-633-2

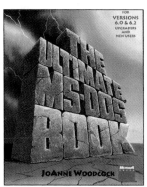

The Ultimate MS-DOS® Book

JoAnne Woodcock

Ever thought you'd chuckle while learning about the MS-DOS operating system? With color illustrations, up-to-the-minute facts, and engaging text, this is the one computer book that will thoroughly entertain you while you're learning about the new MS-DOS 6.2 upgrade. This easy-to-use, informative, and often witty guide will help you get up and running painlessly, so you can quickly take advantage of the new performance and safety features of MS-DOS 6.2.
352 pages, softcover $22.95 ($29.95 Canada)   ISBN 1-55615-627-8

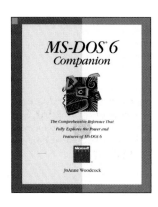

MS-DOS® 6 Companion

JoAnne Woodcock

The comprehensive reference that takes you under the hood of your PC to help you achieve the best performance possible with *all* the features MS-DOS 6 has to offer. This friendly, fact-filled book begins with an overview of MS-DOS and includes information on using the MS-DOS Shell and descriptions of all the MS-DOS commands. Then on to more advanced topics—utilities, batch files, and macros. Includes solid advice and scores of detailed examples.
800 pages, softcover $27.95 ($37.95 Canada)   ISBN 1-55615-550-6

Microsoft Press

Microsoft Press® books are available wherever quality books are sold and through CompuServe's Electronic Mall—GO MSP.
Call 1-800-MSPRESS for direct ordering information or for placing credit card orders.*
Please refer to BBK when placing your order. Prices subject to change.

*In Canada, contact Macmillan Canada, Attn: Microsoft Press Dept., 164 Commander Blvd., Agincourt, Ontario, Canada M1S 3C7, or call (416) 293-8464, ext. 340. Outside the U.S. and Canada, write to International Sales, Microsoft Press, One Microsoft Way, Redmond, WA 98052-6399.

imize Your MS-DOS® Productivity

Microsoft Press® Guide to DoubleSpace™
Doug Lowe

Double your hard disk, work smarter, and change the way you work with MS-DOS 6.0 and 6.2 with this new guide from Microsoft Press. Author Doug Lowe shows you the ins and outs of how DoubleSpace works, how it affects the way you use other MS-DOS commands, and how to use DoubleSpace to its fullest capacities to store more data on your hard disk. You'll learn on how to safely install DoubleSpace, how to use proven strategies for protecting data, how to more than one compressed drive, how to troubleshoot DoubleSpace, and much more.

224 pages, softcover   6 x 9 $14.95 ($19.95 Canada)   ISBN 1-55615-625-1

MS-DOS® to the Max
Dan Gookin

This is the ideal book for users who want to use MS-DOS to make their system scream! In his humorous and straightforward style, bestselling author Dan Gookin packs this book with information about getting the most out of your PC using the MS-DOS 6 utilities. The accompanying disk includes all of the batch files and debug scripts in the book, plus two configuration "Wizards" and several bonus tools that will push your system *to the Max*.

336 pages, softcover with one 3.5-inch disk
$29.95 ($39.95 Canada)   ISBN 1-55615-548-4

Concise Guide to MS-DOS® Batch Files, 3rd ed.
Covers through version 6.2

Kris Jamsa

Batch files offer an easy and instantly rewarding way to significantly increase productivity, without programming experience or additional software! Now updated to cover MS-DOS 6.2, this handy reference book provides intermediate to advanced users information on the fundamentals of batch files, new uses for batch files, and even how to debug batch files.

224 pages, softcover   6 x 9 $12.95   ($16.95 Canada)   ISBN 1-55615-638-3

*Microsoft*Press

icrosoft Press® books are available wherever quality books are sold and through CompuServe's Electronic Mall—GO MSP.
Call 1-800-MSPRESS for direct ordering information or for placing credit card orders.*
Please refer to BBK when placing your order. Prices subject to change.

n Canada, contact Macmillan Canada, Attn: Microsoft Press Dept., 164 Commander Blvd., Agincourt, Ontario, Canada M1S 3C7, or call (416) 293-8464, ext. 340.
Outside the U.S. and Canada, write to International Sales, Microsoft Press, One Microsoft Way, Redmond, WA 98052-6399.

More Easy-to-Use Resources

The Way Computers & MS-DOS® Work
Simon Collin

This is the only book that *shows* you exactly how to use your computer and the MS-DOS operating system. Illustrated in full color throughout, this book shows first-time computer users and beginning users how to set up a PC and perform common computer tasks with ease. It clearly explains the inner workings of the PC, monitor, mouse, and more with the colorful help of detailed illustrations. Features a host of tips, shortcuts and friendly advice from the WYSIWYG Wizard, highlighting the exciting features of MS-DOS 6.0.

128 pages, softcover   8 x 10 $18.95 ($24.95 Canada)   ISBN 1-55615-568-9

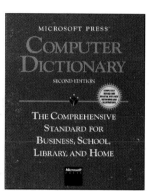

Microsoft Press® Computer Dictionary, 2nd ed.
Microsoft Press

This rich and comprehensive dictionary meets the information needs of today's microcomputer user—ideal for the hundreds of thousands of people who work with microcomputers but who may not be computer professionals. Each of the more than 5,000 entries is written in clear, standard English, and most go beyond simple definition to provide useful detail. Includes pronunciation guides and dozens of new illustrations. The *Microsoft Press Computer Dictionary* was written by a distinguished team of experts from the computer, business, and academic industries.

456 pages, softcover $19.95 ($26.95 Canada)   ISBN 1-55615-597-2

Microsoft® Publisher by Design
Luisa Simone

You don't have to be a professional designer to produce high-quality publications with Microsoft Publisher. This example-driven "how-to" book will help you design logos, business forms, letterheads and letters, advertisements, mail-order catalogs, three-fold brochures, and newsletters. Includes sample projects, power tips, and troubleshooting pointers. Covers version 2.0.

480 pages, softcover $24.95 ($33.95 Canada)   ISBN 1-55615-565-4

Microsoft Press

Microsoft Press® books are available wherever quality books are sold and through CompuServe's Electronic Mall—GO MSP
Call 1-800-MSPRESS for direct ordering information or for placing credit card orders.*
Please refer to BBK when placing your order. Prices subject to change.

*In Canada, contact Macmillan Canada, Attn: Microsoft Press Dept., 164 Commander Blvd., Agincourt, Ontario, Canada M1S 3C7, or call (416) 293-8464, ext. 340
Outside the U.S. and Canada, write to International Sales, Microsoft Press, One Microsoft Way, Redmond, WA 98052-6399.